MATHEMATICS
AS A
CULTURAL SYSTEM

Foundations and Philosophy of Science
and Technology Series

General Editor: MARIO BUNGE, McGill University, Montreal, Canada

Some Titles in the Series

AGASSI, J.
The Philosophy of Technology

ALCOCK, J.
Parapsychology: Science or Magic?

ANGEL, R.
Relativity: The Theory and its Philosophy

BUNGE, M.
The Mind–Body Problem

HATCHER, W.
The Logical Foundations of Mathematics

SIMPSON, G.
Why and How: Some Problems and Methods in Historical Biology

Pergamon Journals of Related Interest

STUDIES IN HISTORY AND PHILOSOPHY OF SCIENCE*

Editor: Gerd Buchdahl, Department of History and Philosophy of Science,
University of Cambridge, England

This journal is designed to encourage complementary approaches to history of science and philosophy of science. Developments in history and philosophy of science have amply illustrated that philosophical discussion requires reference to its historical dimensions and relevant discussions of historical issues can obviously not proceed very far without consideration of critical problems in philosophy. *Studies* publishes detailed philosophical analyses of material in history of the philosophy of science, in methods of historiography and also in philosophy of science treated in developmental dimensions.

*Free specimen copies available on request

MATHEMATICS AS A CULTURAL SYSTEM

by

RAYMOND L. WILDER

University of California, Santa Barbara

PERGAMON PRESS

Oxford · New York · Toronto · Sydney · Paris · Frankfurt

U.K.   Pergamon Press Ltd., Headington Hill Hall,
                        Oxford OX3 0BW, England
U.S.A.   Pergamon Press Inc., Maxwell House, Fairview Park,
                        Elmsford, New York 10523, U.S.A.
CANADA   Pergamon Press Canada Ltd., Suite 104, 150 Consumers
                        Road, Willowdale, Ontario M2 J1P9, Canada
AUSTRALIA   Pergamon Press (Aust.) Pty. Ltd., P.O. Box 544,
                        Potts Point, N.S.W. 2011, Australia
FRANCE   Pergamon Press SARL, 24 rue des Ecoles,
                        75240 Paris, Cedex 05, France
FEDERAL REPUBLIC   Pergamon Press GmbH, 6242 Kronberg-Taunus,
OF GERMANY   Hammerweg 6, Federal Republic of Germany

First edition 1981

British Library Cataloguing in Publication Data

Wilder, Raymond Louis
    Mathematics as a cultural system. — (Foundations
    and philosophy of science and technology series).
    1. Mathematics — Social aspects
    I. Title   II. Series
    303.4'83   QA21   80-41255

ISBN 0-08-025796-8

Printed in Great Britain by
Hazell Watson & Viney Ltd, Aylesbury, Bucks

To my daughter
Beth Dillingham

Introductory Note

There is presented here *a way of looking* at mathematics and its history. Justification for this may be found in any aspect of modern science that has proceeded from the empirical to the theoretical. In dealing with empirical activities, the word "truth" may be aptly used. It is true that certain species of birds fly south in winter and north in summer. But theories explaining such behavior may or may not be true in the same sense. If the behavior seems to conform to a theory T, then T may be accepted as an explanation of the behavior; but to call T "true" in the same sense in which we state observed properties of the behavior would be unjustified.

Similar remarks apply even more forcefully to physical theories — the "big bang" theory, for instance. The classical Newtonian theory of the universe is a good case in point; long considered "true", we know now that it is only a theory, even though a fine instrument with which to work within suitable limitations.

Similarly, the way of looking at mathematics presented herein is not asserted to be the "true" state of affairs. I do assert, however, that to conceive of mathematics as a cultural system does offer a way of explaining many anomalies that, in my opinion, have not been heretofore satisfactorily explained by philosophical or psychological means. Moreover, to consider the evolution of mathematics from a culturological point of view is no more demeaning to the individual mathematician than is the biological theory of evolution demeaning to the individual *Homo sapiens.* Whether it presents a *true* picture or not is not for me — or anyone else — to affirm. That is does constitute a logical and satisfying explanation of mathematical behavior and its history seems to me true — as I hope that some, at least, of my readers will agree.

Also I hope that this work will not be considered as a *history,* such as was the fate of my earlier work *Evolution of Mathematical Concepts* ("EMC"). "Evolution" and "History" are not, I believe, synonymous, although many seem to think so. Historical events are cited in the present work, but only to exemplify or justify the theory; sometimes the same historical event will be cited more than once when its various aspects furnish evidence for more than one theory. It will not be expected, then, that I shall always go to prime sources such as archives, since I shall usually cite the place in the literature where the event will most easily be found by the general reader.

Familiarity with EMC is not assumed in the present work, although sometimes references to it will be made. While EMC may be adumbrative of the present work, the latter is intended as a more mature treatment in that the citations to mathematical theory are not restricted to number and geometry, as in EMC, and concepts (e.g. consolidation, hereditrary stress) which were introduced somewhat superficially in EMC are here analyzed and more explicitly related to mathematical developments. It is not intended, however, that this book is designed only for mathematical readers. Although a particular mathematical concept here and there may be unfamiliar to the general reader, usually its omission will not cause misunderstanding of the general context. Certainly social scientists and philosophers should be able to read and understand. Despite the general misunderstanding of the intent of EMC, it was gratifying that many philosophers and social scientists seem to have discovered and understood the book.

Regarding details of presentation. The Bibliography is organized by author and date; thus Kroeber, 1917:168 refers to the work of Kroeber cited in the Bibliography under the date 1917, and specifically to p. 168 thereof. Since there have been two editions of EMC — a hard cover (Wilder, 1968) and a paperback (Wilder, 1974, 1978) — I shall use "EMC_1, to denote the paperback edition, while "EMC" generally denotes the hardback edition. However, when the reference to EMC and EMC_1 are the same (thus "EMC:Chap 2" and "EMC_1:Chap 2"), the one reference EMC will be used.

Reference to chapters will by by roman numerals. Thus, III-6 refers to section 6 of Chapter III.

In conclusion, I wish to express my thanks to my daughter Professor Beth Dillingham, a professional anthropologist, who encouraged me to write this book and has read the manuscript with a critical eye. Naturally, any errors that I have made are my responsibility, not hers.

R. L. Wilder

University of California
Santa Barbara
June 22, 1980

Contents

CHAPTER I

The Nature of Culture and Cultural Systems

' . . . it appears that the total work of science must be done on a series of levels which the experience of science gradually discovers.'

A. L. Kroeber, 1952: 121

As a group, mathematicians (along with natural scientists) have been generally uninterested, often disdainful, of the social sciences. This attitude has undoubtedly been fostered, at least to some extent, by the appearance now and then of an article purporting to prove a social fact which seems already "obvious" to any perceptive observer of the current social scene.

On deeper reflection, however, it seems to be true that the mathematician is guilty of taking too much as "obvious" in the realm of human behavior. As a human being possessed of unusual talent, he is quite aware of his own feelings and reactions to his social environment, and it is easy for him to conclude that his own perception thereof is perfectly clear and unneedful of explanation on behavioral grounds.

This attitude can be defended, to be sure. Contrary to much general opinion, the modern mathematician is not just a narrow specialist unaware of the intellectual life about him. His love of, and oftentimes skill in, music is generally known; the question of the relationship between music and mathematics on the plane of interest and appreciation has for long been an enticing one, never satisfactorally explained. Love of the arts generally, and interest in the humanities as well as social and political affairs, are common among mathematicians. Usually scornful of display, and often lacking in forensic ability, the mathematician has been content to hide his extracurricular talents even from his fellows in mathematics. Only in mathematics itself does he depart, as a rule, from this custom, and here his creativity he does not seek to conceal.

It is all the more unfortunate, then, that he has often allowed his sense of superiority in intellectual affairs to close his mind to the more remarkable advances in the social sciences. And of the latter, hardly any is more

1

remarkable than the discovery of the concept of culture. Not only has this cut the Gordian knot of problems surrounding differing races and nations, but it can serve as a tool of research in all areas of science, natural as well as social. Along with the concept of evolution, it can clarify the distinction between *Homo sapiens* and the other forms of life. Even where the presence of elementary (but questionable, nevertheless) forms of "culture" have been asserted as present among primates, the human form of culture is distinguished by the capacity for evolution. In the short space of a few millenia, culture as possessed by *Homo sapiens* has evolved from the types of culture found among primitive hunting tribes to those found in civilized societies today (we can only guess at the degree of evolution which preceded).

Of perhaps greater importance to the mathematician, however, is the contribution that an appreciation of the impact of his culture upon him and his work, can make. And, as I have already tried to point out in earlier publications,[1] such knowledge, substituting for vague intuition, can profoundly influence his choice of problem, as well as his attitude toward the works of his fellow workers.

Consider, for example, the attempts made by Russell and others to base mathematics on logic. Underneath all of these early researches was the basic assumption that the logic they were analyzing for their purposes was an indisputable absolute. Only later, when different "logics" were discovered by both anthropologists and mathematical logicians, did it become clear that their logic was a cultural artifact, not a necessary component of every culture.

The original choice of the term "culture" was perhaps unfortunate. It has too many connotations. When it appears in news items — as in the term "European culture," for instance — it is usually questionable whether the author really sensed the significance of the term. In such cases, the term "society" would usually convey the meaning of the author. Then of course there is its use in such contexts as "cultured gentleman," not to mention other uses such as "bacterial culture." As a result, one must ordinarily rely on the context in which the term is used to make its meaning clear. But is it ever clear? It appears that any author making extended use of the term can avoid being misunderstood only by making explicit what he means by the

---

[1]See, for example, Wilder, 1953, pp. 425, 439, 445.

term. This is true even for anthropologists, for whom it denotes a basic notion, since the actual character of "culture" is not at all generally agreed upon even by them.

It is all the more important, then, that *anyone who writes on topics requiring an understanding of the concept of culture, should make a special effort to indicate exactly what he means thereby.* I have not only learned this by experience, but have also learned that brief, casual explanations are insufficient.

1. Evolution of a cultural artifact.

Instead of immediately stating a definition of culture, let us begin by considering an example of how a part of a culture, specifically a *cultural element,* might evolve.

Let P_1 denote a person who has come across a bird, *B,* which interests him enough that he wishes to study it in depth. We suppose P_1 has never seen such an object as *B* before, although this particular *B* allows almost unlimited observation of itself and its activities. Because of his interest in *B,* P_1 begins to form certain opinions about *B* and to reach certain conclusions regarding the nature of *B.* The cluster of these opinions and conclusions, as they are ultimately developed by P_1, we denote by Q_1.

Let P_2 denote another person, unknown to P_1, who also observes *B* and forms an analogous collection, Q_2, of opinions and conclusions about *B.* We then have the pattern sketched in Fig. 1, where the lines denote the inflow of opinions, etc., resulting from observations of *B* by P_1 and P_2.

We next suppose that P_1 and P_2 become acquainted, and that they reveal to one another that they have been observing *B,* and that they disclose to one another the contents of Q_1 and Q_2. The new situation may be sketched as in Fig. 2. The stores of opinion Q_1 and Q_2 are represented as intersecting; i.e. certain opinions about *B* arrived at by P_1 and P_2 turn out to be identical (represented by the shaded area). It is possible, of course, that none will turn out to be identical or, on the other hand, that they coincide throughout.

It is also possible that after some discussion by P_1 and P_2, the two stores of opinion Q_1 and Q_2 come to have more in common, or, eventually, to coincide (complete agreement). On the basis of rational discussion, common agreement regarding the physical aspects (color, song, etc.) may

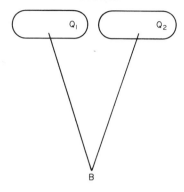

Fig. 1

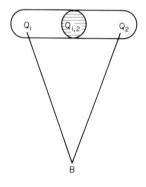

Fig. 2

be reached, while P_1 and P_2 may each have added to the common store of opinion by observations made by one and not the other. In any case, let us denote by $Q_{1,2}$ their area of common agreement.

It it important to recognize, at this point, another feature of the relationship between $Q_{1,2}$ and P_1. As part of Q_1, the collection $Q_{1,2}$ is no longer the exlusive property of P_1; he now shares it with P_2. A similar remark holds regarding the relationship of P_2 and $Q_{1,2}$.

Continuing, we may bring other persons, P_3, P_4, ..., P_i into the picture.

Assuming they enter one at a time, we denote the area of their common agreement concerning B by $Q_{1, 2, ..., i}$. At some stage, the necessity for new terms may arise, presumably agreed upon by the persons concerned and denoting special aspects of B and its habits. This will undoubtedly be necessary when the interest in B grows to the stage that publication of the conclusions of the various observers P_i becomes desirable. At this point, discussion concerning B proceeds largely on the impersonal level of journal articles, until either the problems raised by B are substantially settled, and publication ceases, or these problems become generalized in the study of a whole phylum. In either case, there will have arisen a whole consensus of opinion, Q (this will presumably be some $Q_{1, 2, ..., i}$); and as time goes by, Q will become a part of the common heritage of those who have a scientific interest in B. In short, Q will now have become a cultural artifact, having evolved from $Q_1, Q_{1, 2}, ..., Q_{1, 2, ..., i, ...}$.

We now observe that Q is no longer just the private opinion of any one of the investigators P_i concerning B's nature and behavior. It is more inflexibly removed from manipulation by any individual observer P_i, having gained a status which precludes alterations or deletions by any individual investigator P_i without the common consent of other investigators. Moreover, as time passes, with the eventual disappearance of the original investigators from the living scene, Q takes on a "traditional" character, forming a store of accepted information about the successors of B, to which are referred all ornithological investigators of later generations. In short, it is then a true cultural artifact. This does not mean that it cannot be altered or augmented by new items concerning type B birds; but doing so now requires submission to the ornithological group concerned and its eventual acceptance by this group; after which it assumes cultural status.

Now the above is obviously over-simplified, and does not adequately represent the manner in which cultural elements generally evolve. For example, P_1 may be a mathematician and B a new concept which P_1 has invented. In the modern scheme of things, P_1 may decide to publish the details of B. Assuming such publication takes place, the concept B may then find acceptance by the particular group, G, of mathematicians to whom it is of interest. Such acceptance usually means that B finds a place in the system which is being developed by the group G, and hence achieves the status of a cultural artifact.

The case of the bird B might have developed in an analogous manner.

Assuming that P_1 is a modern ornithologist, P_1 might publish his set of opinions Q_1 so that the ornithological community would be made aware of them almost immediately. In this case, P_2, P_3 and other ornithologists who also observe B would check the details of Q_1 against their own observations, and the discussion concerning the nature of B would then occur in ornithological periodicals. Ultimately, the consensus, Q, would as before become part of the standard ornithological materials of the day.

2. The things that make up a culture

But a culture of a given society (e.g. ethnic or national) does not consist of just the findings of ornithologists, mathematicians or other scientists. To an "average" person, P, in a modern society, his "culture" consists of the general collection, C, of beliefs, prejudices, ways of dealing with others, and the knowledge required to do his job, whatever it may be. Unlike the acquisition of new knowledge (as in the case of the ornithologist and mathematician described above), this collection, C, was not invented by P or discovered by him; most of it was in existence before he was born. He acquired it through his contacts with others, via a communication process. As a child, he learned from his elders how he was expected to behave and, moreover, what he was to believe religionwise. At the latter stage he also learned how to count, through a "pointing" process used by his elders. What are called "values" were acquired in similar fashion; and the entire accumulation of beliefs, attitudes and general knowledge thus acquired came to constitute for P "what is right and what is true." In short, what we may call his world view, W, is thus established, and unless he approaches life thereafter with an open mind, willing to assess and possibly accept concepts and values that are inconsistent with W, he will abide by the principles of W for the rest of his life (which seems to be the case with the majority of people).

It is the latter fact that makes life difficult for advocates of social change. Currently (1980) in the United States there is strong support for equality of men and women, especially for equal wages, and even to the extent of participation in the armed forces. However, parents, yielding to cultural tradition, instil different world views in male and female children and purposely arrange that when their children reach adulthood they shall conform to the cultural *mores* regarding differences in the sexes. As a consequence, achievement of real equality is rendered difficult. Similar remarks can be made of religious beliefs which, due to their inculcation at an early age, achieve a firm resistance to change. Early mathematical training, too, is usually carried out in such a way as to affirm the absolute truth of mathematics; herein lies the source of much platonism, no doubt.

In EMC I defined a culture as "a collection of customs, rituals, beliefs, tools, *mores,* and so on, called *cultural elements,* possessed by a group of people who are related by some associative factor (or factors) such as common membership in a primitive tribe, geographical contiguity, or common occupation." I should have included "language" among the cultural elements, especially since it forms the cement which binds together the world views of a people. Let us expand upon this in more detail.

Suppose, for example, that P denotes the group of people having a given nationality, say the group of all Chinese people. Just what is their culture? When we say that it is the "collection of customs, rituals, beliefs, etc. ... " of these people, do we mean to include the beliefs of a particular Chinese gentleman, G, even if his beliefs do not coincide with the beliefs of the majority? The answer is yes; for as long as he lives, his beliefs will constitute a portion of Chinese culture, "a constellation of cultural elements, a capsule of culture" (White, 1975: 9).

To make this clearer, suppose I, who have never visited China, make a visit thereto. Presumably I will meet a certain number of Chinese people, from whom I will gather certain concepts or elements of the Chinese culture. These will be augmented by my observing how people in general act in public — on the streets, in religious gatherings, etc.; also by reading newspapers (assuming I have picked up enough knowledge of the language), books in libraries, etc. Notice, however, that all of my contacts, even the books in libraries, are derived from the behavior and beliefs of certain individuals. And if I should decide to become (and am accepted as) a Chinese citizen, my own concepts and beliefs will constitute a part of the Chinese culture. This is true even though my world view, W, was originally derived from my upbringing in the United States. Of course, it is equally true that a person of Chinese descent who becomes a naturalized U.S. citizen will bear a like relationship to the U.S. culture; his world view will constitute a part thereof.

Yet, and this is important, *no individual person in a group will possess all the culture of that group,* even though all group members may seem to "think alike." This is trivially true, since individual differences due both to heredity and early environment preclude it; no two individuals are the same in their world views (even though a particular two may seem to be). The culture of the group, to repeat, consists of the *totality* of the individual world views, even those who may be dubbed "mentally unbalanced," *united by bonds of communication.*

3. Culture as a collection of element in a communications network

A culture is not just a *collection,* however, any more than a collection of gears, wheels, etc., constitutes an engine. In the latter case each gear in the engine is related to the totality of other parts of the machine in a definite fashion, and contributes to the thing we call "engine;" this relationship is expressed in the general plan of the engine and operates through the *contacts* made by the gear with other machine components. In the case of a culture, there is likewise a mode of relationship, which operates between the individuals of the culture, namely *communication.*

Communication in a culture is based on *symbols* — ultimately on the *symbolic faculty* possessed by *Homo sapiens.* This faculty enables its possessor to exercise a *naming* function; in particular if a new object never before observed is encountered, it is given a name, and the name is thereafter used to indicate this particular object. This process goes on all the time in mathematics, and more generally in science and industry. Articles with new names assigned are all the time entering the market. Similar remarks hold for non-material items such as concepts, and we are enabled thereby to indulge in abstract thinking. Symbols occur not only in the form of names; the figures of geometry — triangles, squares, etc. — are symbols; they are particularly important for so-called visual thinkers.

But most of the names that we use are not invented by us; they were assigned to their designatees long before we were born. For us, they are *signs,* and they are most important; how could we exist as a society if we continually exercised our symbolic faculties to give new names to things from one generation to another, for instance? To be sure, over the long term, names and all words change, but they do so very gradually so that a given generation does not notice the change. The changes are not usually initiated by individuals, but generally occur unnoticed; witness how English has evolved from Old English in a few centuries, and how languages such as French, Spanish and Italian grew out of the classical Latin in a span of only two millenia.

The representation of languages in the form of written symbols has enabled philologists and linguists to trace the evolution of languages over many centuries. Languages cannot remain stable, since by its very nature, a language is a device for communication between *individuals*. As in the case of whole cultures, contacts inevitably lead to language change, even within a single culture (although they are accelerated by contacts between cultures). The rapidity of change in a single language, English, is well illustrated by the necessity for dictionary-makers to decide to what historical period of time they will limit their choice of words — a tacit recognition of linguistic evolution. In the case of the Oxford dictionary,[2] this limit was extended from A.D. 1150 to the present.

It is worthwhile quoting from the introductory "General Explanations" of this work:

> The living vocabulary is no more permanent in its constitution than definite in its extent. It is not to-day what it was a century ago, still less what it will be a century hence (an indication of the accelerated tempo of change due to increased ease of communication — R.L.W.). Its constituent elements are in a state of slow but incessant dissolution and renovation. "Old Words" are ever becoming obsolete and dying out: "new words" are continually pressing in. ... the vast majority of the ancient (prior to 1150) words that were destined not to live into modern English, comprising the entire scientific, philosophical, and poetical vocabulary of Old English, had already (by 1150) disappeared, and the old inflexional and grammatical system had been levelled to one so essentially modern as to require no special treatment in the Dictionary. Hence we exclude all words that had become obsolete by 1150.

With the example of language, the basis of communication in a culture, we can see more clearly how a culture is constituted. The portion of culture (beliefs, *mores,* language, etc.), possessed by an individual, is continually linked to the portions possessed by others through the medium of communication. The individual cannot make any radical changes in the culture without cooperation from his associates. But *the changes do occur;* language changes are only a part of the changes that a culture undergoes. And since they are usually not initiated on the individual level, we must

---

[2]Our references here are to the Compact Edition of the *Oxford English Dictionary,* Oxford University Press, 1971; see p. x.

treat them as occurring on a super-organic level which we now call "culture."

This last sentence requires explanation. Did not Marconi, an *individual,* invent wireless communication? Did not the Wright brothers invent the airplane? Did not Alexander Graham Bell invent the telephone? These events certainly initiated cultural changes of vast importance.

In the case of Marconi, what about Maxwell and Hertz, for instance? In the Wrights' case, what about Otto Lilienthal and Langley? It took a famous lawsuit to decide that Bell invented the telephone?[3] In each of these cases, the so-called "inventor" took a critical step in a series of steps leading to his invention; he was totally dependent not only upon ideas he gleaned from others, but, more important, for the *push* to invent which already existed in his culture. Actually, inventions are *collective;* i.e. cultural achievements. The inventor assembles all the ideas of his predecessors and, oftentimes, his contemporaries. True, he takes a critical step; but this is a subjective judgement, since all the steps leading to his invention were just as critical, in the sense of being necessary to the final invention, as the latter itself.

The question of the relationship between an individual and his culture has found many differing answers among anthropologists. Some have declared that the individual *makes* his culture (referring to the general culture in which he participates); others, that the culture makes him! The latter opinion stems from recognition of the part that culture plays in determining for an individual his values, beliefs, etc., i.e. his *world view.* The former opinion derives from the fact that new concepts, inventions, etc., usually are devised and contributed to the culture by individuals.

We encounter here the "great man" concept of cultural evolution. According to this concept, a culture advances because of the inventions (conceptual, mechanical, etc.) of its great men and the presence of the great man at any given time is a genetic accident. Unfortunately for this theory, it is well recognized now that potentially "great men" are always present. "Nothing now known in biological heredity, nothing in the laws of chance, can account for the tremendous variations in the frequency of genius. The only explanation yet advanced which is not wholly speculative or arbitrary

---

[3]"The ultimate decision rested on an interval of hours between the recording of concurrent descriptions by Alexander Bell and Elisha Gray" — Kroeber, 1917: 200.

sees a correlation between realized genius and opportunity given by the stage of a civilization's development — the stage where the productive cultural patterns are defined and mature but where their inherent potentialities have not yet begun to be exhausted" (Kroeber, 1952: 128). "What we are wont to call 'great men' are those among many more individuals of above-average ability who happen to get born in a time and place and society the patterns of whose culture have formed with sufficient potential value and have developed to sufficient ripeness to allow the full capacities of these individuals to be realized and expressed" (Kroeber, *loc. cit.*).

In mathematics, for example, when a new field of mathematics is just opening up, it is a particularly opportune time for the young Ph.D. to become involved in it. It is sometimes said that, when advising a new graduate regarding his choice of a graduate institution, he should first select a mathematician known to be an inspiring teacher with a creditable list of doctorates, and then enroll in his courses; but it should be added that the *state of the field* in which such a man works should also be taken into account. His capacity for inspiring students may possibly be over-weighted by the fact that his field of interest is becoming pretty well worked out.[4]

As to the *existence* of culture, there is general agreement, in much the same way as scientists may agree regarding the existence of a natural phenomenon without, however, agreeing upon its nature. From a *scientific* point of view, it is preferable to adopt a *theory* of culture which best lends itself to explanation and prediction of cultural events. No science ever succeeds in establishing a basic philosophy which can be treated as once and for all "explaining" its phenomena. Witness not only the natural sciences — physics, chemistry, zoology — but mathematics. The non-mathematician who thinks that mathematics rests on a secure base of "truth" is simply unaware of the shifting sands that underlie all of mathematics. Mathematicians no more agree on the nature of mathematics than do anthropologists in regard to culture. Nevertheless, mathematics "works" in all kinds of applications, and similarly, the concept of culture "works" as both an explanatory and predictive device in the social sciences.

Physical scientists may not *know* what constitutes a physical entity; but

---

[4]Of course, fields in such a state can frequently be rejuvenated. This will be discussed later, particularly when we come to the concept of *consolidation* (Chapter V).

they can observe its effects and erect theories about its nature which will have tremendous consequences. In particular, no one doubts the entity's "existence," or dubs a theory explaining it as "metaphysical." We should not, therefore, be deluded, because of its intimate relation to our actions and beliefs, into denying the existence of culture, condemning as "metaphysical" a theory of culture which treats it as something superorganic whose influence on our beliefs, actions and manner of speaking is all too evident. That "something" which, at a particular period of time, prevented white American males from wearing earrings, but made it proper for an American Indian to do so, we call *culture*; it seems to be changing so that white American males are starting to wear earrings (on *one* ear). We do know that we cannot individually change such customs; that any attempt to do so on our parts will result in ridicule and suspicion of our sanity. I also know why I, as an individual, entertain certain beliefs and observe certain customs; the culture in which I was reared and lived as an adult both dictated, and later afforded means, whereby I was enabled to make some individual changes in *my* beliefs (e.g. regarding the nature of mathematics).

Where do speakers get the words they speak; and where do inventors get the ideas whose consolidation leads to their inventions? The answer is, "in the already existing culture whose carriers they are." If asked to describe the culture whose elements he has utilized, the speaker and/or inventor can hardly enumerate what this individual, that individual, etc., knew and believed — he must use the *collective* concept, that which embraces all the individual elements, to answer this question. Moreover, his conception of this complex of knowledge, books, etc., that have formed the basis for his work, is usually as a collective of unspecified elements.

The bulk of our beliefs, *mores,* and technology, have congealed into a heterogenious mass of elements whose origins are lost in unwritten history. Even in the case of mathematics, where written histories are available for the more recent inventions, there exists a "tradition" composed of logic,[5] mathematical folklore, unwritten rules regarding priority rights, publication modes (which have changed over the centuries along with the invention of new techniques of reproduction), rewards systems, etc. Moreover, since the average mathematician knows little or no history of his

---

[5]We refer here to the logic used by the working mathematician, not formal or mathematical logic; i.e. that logic which is subsumed under general modes of proof in common use.

subject, a kind of folklore must be added to his store of tradition. This tradition, along with the mathematics itself, of course, forms his mathematical culture as a subculture of the general culture in which the mathematician lives. Its controlling character is just as real as that of the general culture, and operates as a super-organic entity quite as efficiently as does the general culture.

It is not only practical, but theoretically sound, to treat culture as a super-organic entity which, like its subset language, evolves according to its own "laws." The final judgement regarding a theory of culture should not be whether it suits one's *beliefs* regarding reality, but whether it works best as an explanatory and/or predictive device.[6] Even if an individual can do nothing about changing the direction taken by a culture, he can *believe* that he can; if it turns out that he is only acting at the whim of his culture, he still may achieve his purpose. History is full of examples where an individual has influenced the course taken by a culture; but analysis of his motivation has shown him to be a catalyst of cultural determinants.

As an adult individual, I experience culture chiefly through reading newspapers, journals and books, but also, and most important, through the reactions I receive — and come to expect — in my dealings with other individuals. Each such individual, I expect, will have certain "pre-conceived" ideas and principles. If I could not expect this, and if nobody else could, then the result would be chaos. Lack of appreciation of this was a contributing cause of the treatment accorded "primitives" by conquering peoples. The latter expected the primitive to observe certain rules (ethical, moral and religious) and to react in certain ways. When such proved not to be the case, violence could be expected to result.

It must be pointed out, however, that mathematics could not always have been considered as a subculture of the general culture. Thus, in Babylon and Egypt, mathematics had evidently not gained a status that would warrant calling it a "subculture;" it was only a *cultural element.*

The problem of when a cultural element can be considered as itself a culture — i.e. subculture of the general culture — is a difficult one — one that has not been treated in the anthropoligical literature to any extent, so

---

[6]Consider, for example, the Axiom of Choice. Now that it is known to be independent among generally accepted axioms of set theory, the decision regarding its acceptance or rejection should depend on what it accomplishes when it is assumed, rather that on any philosophical basis.

far as I know. Generally, a social organization such as a women's club, or the alumni of a university, would not qualify as subcultures. But would the Masonic order, for example, merit being called a "subculture?" Or the Democratic Party in the United States? These questions are beyond the scope of the present study, as is, moreover, the question of *when* mathematics should be considered as having passed from being only a cultural element in a given culture to being a subculture of that culture. That mathematics today constitutes a subculture of the general culture seems generally agreed upon. As remarked above, it has its own traditions and, its own "laws" of development. Probably the criteria as to whether a given cultural element may be considered a subculture should at least contain these provisos: that it has its own traditions, clearly and uniquely identifiable within the traditions of the general culture, and its own laws of development (which may or may not coincide, in part at least, with laws of development of the general culture).

Another feature of culture which must be emphasized is its "time-binding" character. A distinctive difference between *Homo sapiens* and other life forms is that in the former knowledge is *cumulative*. A new generation does not have to re-do or re-invent concepts which were created by the older generation. Rather, the new generation takes up from where the older left off, and goes on from there. In this way, a culture grows, or *evolves*.

4. Mathematics as a cultural system

This is a further theoretical device, adapted with revisions from the late anthropologist L. A. White's work (White, 1975).

By way of introducing it, we recall that a favorite picture which has been in use in various articles and treatises for many years to represent graphically the structure of mathematics, is the so-called "Tree of Mathematics." According to this conception, mathematics can be represented as a tree-like configuration whose roots are the Foundations of Mathematics and whose trunk and branches represent the various subfields of mathematics. Where a subfield, such as graph theory, splits off from its parent, topology, it is represented accordingly by a branch of the tree growing out of the larger branch representing topology.

This was very useful so long as mathematics grew this way — thus algebra and geometry were distinct branches, and the theories of axiomatics and logic were represented as roots. But as long ago as the 17th century when Fermat and Descartes introduced their consolidation of algebra and geometry to form analytic geometry, the "tree" representation began to break down. For here branches of algebra and geometry came together again to form a new branch of mathematics, analytic geometry. And later, work in the foundations of mathematical logic not only brought about consolidation between branches and roots — e.g. algebra á la Boole — but brought the roots up to aid in the solution of problems belonging in the branches (e.g. the continuum problem of the theory of sets, Souslin's problem, etc.). Such events not only made the tree analogy quite unrepresentative of the way mathematics grows, but emphasized the interrelatedness of all parts of mathematics — foundations as well as the most advanced topological theories, for example.

Of course, any manner of looking at the growth of mathematics is going to run into such difficulties as assigning particular mathematical theorems to their proper niches. For instance, is Euler's polyhedral theorem a theorem of classical geometry — which it was considered originally to be — or a theorem of topology, to which it is usually now assigned? Any system of representation will have to allow considerable flexibility. White's theory of cultural systems, as we shall adapt it, accomplishes this flexibility.

In White's theory of the representation of a culture as a cultural system, or subdivision thereof, the entity is conceived as a system of vectors, where each vector represents a particular type of interest, e.g. the agricultural segment of the culture, or the religious portion. Other vectors might represent oil interests, manufacturing interests, educational interests, etc. Each of these interests, depending upon the situation under consideration, can be split into vectors if necessary. For example, each particular religious sect might be considered a vector, instead of representing the totality of religious interests by one vector. In the dispute over the introduction of prayer into the public schools, the religious lobby would preferably be considered as one vector, while for other purposes, each religion might be treated as a vector.

Each vector, as in the use of vectors in mechanics, has both magnitude and direction; magnitude can be measured in terms suitable to the problem in question, such as the number of people involved, amount of money at its disposal, etc., while direction may be conceived in a manner appropriate to the issues involved — in some cases a simple pro or con (+ or −).

Now in our culture, one of the possible vectors is *science,* both pure and applied. It is this vector that is referred to when there occur in the public media such statements as "Science shows thus and so," or "Science is the major influence in the advance of civilization." But then for some purposes we can split science into vectors consisting of physics, chemistry, anthropology, psychology, mathematics, and so on. (Incidentally, the departments in an academic community can be considered as vectors, and the operations of these vectors observed in any faculty meeting where academic issues are being decided.)

For our purposes, we consider mathematics as a subdivision of our general culture, and instead of representing it as a tree, think of it as a system of vectors. Thus geometry would constitute one vector, algebra another, topology another, and so on; or we could base our system of vectors on the subjects given in the journal *Mathematical Reviews'* classification of the mathematical sciences. The picture I wish to draw is of a vectorial system, mathematics, in which each vector is striving for further growth and in which the different vectors impinge on one another, offering assistance by diffusion of ideas to other vectors, sometimes resulting in new consolidations which will become vectors in their own right. The foundations vector is no longer subordinated to being "roots" of a tree, but is a vector among other vectors operating in much the same way as these other vectors.

The assignment of vectors (i.e. vector status) will depend on the problem to be studied. For my present purposes, the assignment of vectors will be governed by eras chronologically ordered; and I will consider the evolution of mathematics as the sequence of its vectorial forms throughout history. I have to skip details, especially regarding the early history of mathematics, and confine myself in the main to medieval and modern history. Then we can conceive of the evolution of mathematics as an ordered array of vectorial systems in which the vectors grow in magnitude at varying rates; in one era the geometric vector grows rapidly while the others remain virtually static; in another era the vector representing analysis begins to accelerate its growth; and in the modern era the theory of sets vector splits off from analysis; and so on.

Incidentally, the same kind of theoretical structure can be applied to other sciences, with similar remarks.

Each vector generates its own stresses or forces, as well as being subject to external stress, both from other vectors and from the outside culture. I have already mentioned diffusion, in which properties of one or more vectors diffuse to another. An important internal stress in mathematics is one that Kroeber called "potential," but which I prefer to call *hereditary stress* since we inherit it from our forebears. Actually, it has many components. Conceptual stress is one such component; for instance, when complex numbers "reared their ugly heads," they were not considered reputable, but as they insisted on intruding into respectable mathematics during the course of time, a niche was finally found for them — they literally forced themselves in. Status is another aspect of hereditary stress; as a branch of mathematics grows, it may acquire such status that everyone, including graduate students, may be tempted to get on the bandwagon. These notions will be treated in more detail in later chapters.[7]

5.   Cultural and conceptual evolution

That most cultures, especially modern cultures, undergo change as time passes should be obvious to every observer of the changes in cultural habits and *mores.* The perennial objections of the older members of a society to the "new ways" of the younger generation bear testimony to the fact of change. Changes within the modern scientific cultures, and in particular mathematics, are constantly before our eyes in the number and contents of research journals. Such changes contribute to what we call *evolution of culture.*

It is desirable to distinguish between particular instances of change and changes in the forms and patterns of cultures. To put the matter in general terms, we should distinguish between *history,* as generally understood, and *evolution.* For example, one would speak of the "history of the reign of King Henry the Eighth," but hardly of the "evolution of the reign of King Henry the Eighth." One could, of course, speak of the "evolution of the state as a form of government," using as evidence special examples from

---

[7]An alternative approach to that which we take, which is qualitative in nature, utilizes so-called "general systems theory," which is already being applied by some anthropologists to the study of cultural models. How successful this approach will be seems undecided. See, for instance, a panel report, Rodin *et al.,* 1978.

history. As the terms have been customarily used, history is a *particularizing* process, whereas evolution is a *generalizing* process.

More precisely, *history* is a *record* — a record of past events arranged in chronological order, together with some discussion of relations (e.g. causal) between these events. On the other hand, *evolution* is a *process of change* — a process by which various forms and structures change into ("improved") forms or structures, and is generally motivated by certain forces whose nature is dependent upon the types of forms or structures involved.

By the term "cultural history" we would mean a record of particular cultural events, whereas by "cultural evolution" we would understand the changes which the forms of a culture or cultures undergo. As a culture, mathematics has been treated chiefly from an historical standpoint. However, it has had an evolution, something which we tried to stress in EMC; and that evolution has been subject to various forces or stresses just as has biological evolution. The evolutionary aspect of mathematical development has been generally neglected.

We do not mean, however, that the two, history and evolution, must be sharply separated, so long as the distinction is maintained. In particular, when one describes special types of change, as, for instance, in the historical development of the calculus, it may be illuminating to point out those aspects of the development that are responses to, or caused by, recognized evolutionary forces, or which fall into a recognized pattern of the evolutionary process. In this way, bringing in evolutionary ideas acts to complement and extend history as it has ordinarily been written.

The anthropologist H. S. Shapiro has illustrated this quite clearly in a discussion of the contribution that the notion of culture can make to history (Shapiro, 1970: Chap. 2). For instance, the reactions of a people or culture to prolonged severe oppression at the hands of a militarily and numerically stronger people has usually conformed to a classical pattern, namely an intense revival of the traditional religion. His example of the long repression of Ireland by England, which resulted in a remarkable revival of Catholicism in that country (maintained to this day), can be augmented by the case of the long-term retention of their faith by the Jews throughout centuries of oppression, and, on a smaller scale, the case of the intense loyalty to their traditional forms of religion by the various Pueblo tribes of southwestern United States after the Spanish conquest.

The recognition of such patterns adds measurably to the understanding of such historical developments, as well as furnishing materials for forecasting the future results of similar events.[8]

Two aspects of cultural evolution will be of especial concern to us in the sequel: (1) those general patterns of change that are exemplified by the work of Shapiro cited above and which have also been exemplified to some extent by so-called "laws" in the final chapter of EMC, by Crowe (1975), Dieudonné (1975) and Koppelman (1975); (2) those detectable stresses or forces that serve to implement evolutionary changes, such as diffusion, hereditary stress, etc. (see EMC).

In mathematics, however, we have to deal with the evolution of *concepts* as well as with the history and evolution of mathematics as a cultural entity. Generally, by the evolution of a concept, we shall mean the development of forms, especially the changes induced in these in response to various needs, which finally resulted in the definition of the concept as currently understood. Consider, for example, the concept of "function" in mathematics. This took a number of forms before the modern form was agreed upon, some of which were the result of arguments between different "schools of thought," and others of which were dictated by the necessity of generalizing in order to meet certain demands. The sum total of these changes, together with the reasons therefore, should properly be called "evolution" of the concept of funtion. We then restrict the term "history" to the record of the actual events that occurred during the "evolution," such as d'Alembert's, Euler's and John Bernoulli's discussions of the vibrating-string problem, trigonometric series and the work of Fourier, and on through the contributions of Hankel, Dirichlet, Riemann, Weierstrass and others.

Our use of the term "evolution" with respect to the development of concepts seems to agree with the use of the term when applied to particular species in biology. For example, the development of the horse through its *evolutionary* stages and the development through its *historical* stages are usually treated separately (e.g. *Enc. Brit.,* 11th ed. and 14th ed.). The evolutionary stages are devoted to the successive forms taken by the

---

[8]Shapiro also analyzed the development of the constitutional form of the United States government in terms of the cultural reactions of a splitting off of subcultures from parent cultures, as in the form of colonialism. His explanation of the early lag in the development of native forms of art and science in the United States is especially illuminating.

"horse" from the earliest known precursor types to the modern horse. The history of the horse is usually devoted to the wanderings and settlements of the modern horse in particular parts of the world, as well as the uses made by man of the horse. Similar distinctions are made between the evolution of man and the history of man. It should be remarked, however, that the idea of evolution of culture did not arise from the biological use of the notion. Indeed, the original uses of evolutionary ideas were to social and cultural theories, as, for instance, in the work of Herbert Spencer (see White, 1959).

CHAPTER II

Examples of Cultural Patterns Observable in the Evolution of Mathematics

Nothing from Nothing ever yet was Born
Lucretius, *De Rerum Natura*

One of the best forms of evidence for the utility of the concept of culture as a super-organic entity, having its own laws of development, and not subject to the whims of the individual, is the types of predictable regularities which can be observed in mathematical evolution. We have already commented upon the similar phenomenon in the case of language in Chapter I, and in § 5 thereof described certain regularities brought out by H. Shapiro in the history of Ireland and the U.S.

In this chapter we give some examples, in the fields of mathematics and science, of cultural patterns characteristic of their evolution. The regularities exhibited therein as well as the dominance displayed by such established patterns of behavior, afford convincing evidence of the workings of a cultural system that can hardly be explained by appeal to chance. We conclude the chapter (§ 12) with a problem in which the notion of culture appears to contain the solution, but in which it is only vaguely determinate how the culture operates.

Partly because the phenomenon has not generally been treated from the standpoint of mathematics, partly for the sake of completeness, and partly for the benefit of readers who may not be familiar with the phenomenon,[1] we begin with a brief discussion of multiples — i.e. cases of multiple discoveries or inventions. They have been noted by discerning scholars for centuries, and occur in all fields of knowledge, as well as having caused

---

[1]For an interesting popular discussion of multiples, see Whyte, 1950. For a more technical and definitive discussion of the phenomenon, see Merton, 1973: Part 4; also White, 1949: 205–211.

innumerable cases of bickering over priority and accusations of plagiarism.[2]

1. Multiples

The classic case of multiples in mathematics was that of Leibniz (1676) and Newton (1671). The acrimonious debate — largely instigated by friends rather than by the principals concerned — that followed is too well known to warrant repeating here. The reader can consult any good history of mathematics for details.

Probably as well known is the Bolyai (1826–1833)–Gauss (1829?)–Lobachewski (1836–1840?) invention of non-euclidean geometry.[3] One of the chief interests attaching to this case is the long period — *circa* twenty centuries — that lapsed between the proposal of the problem by the Greeks and its ultimate solution. This, together with Saccheri's failure to solve the problem, can be explained satisfactorily by culturological means. Certainly the "clustering of genius" in the early 19th century cannot be attributed to chance!

Probably less well known are such early cases as the invention of logarithms by Napier-Briggs (1614) and Bürgi (1620); discovery of the principle of least squares by Legendre (1806) and Gauss (1809); the geometric law of duality by Plücker, Poncelet and Gergonne (early 19th century).

These can be supplemented by many other cases which have occurred in the various subfields of mathematics especially in modern times and are best known by the workers therein. For example, topologists familiar with the history of their field know of the famous Hahn–Mazurkiewicz multiple which occurred in 1914, at a time when communication between scientists in warring nations practically ceased. This multiple is of special interest in

[2]In his classic paper (1917), A. L. Kroeber remarks of the discovery of anesthetics in 1845–6 by no less than four men: "So independent were their achievements, so similar even in details and so closely contemporaneous, that polemics, lawsuits and political agitation ensued for years, and there was not one of the four but whose career was embittered, if not ruined, by the animosities arising from the indistinguishability of the priority" (*ibid.,* p. 200).

[3]For a concise discussion of this case, see Coolidge, 1940: 72ff. Coolidge remarks, "It is one of the surprising facts in mathematical history that at various times, important results have been discussed independently by different men." On a cultural basis there is, of course, nothing surprising about it.

that despite the overriding importance of the political situation, the mathematical subculture continued to develop in the direction that was plainly indicated by preceding events.[4] Similarly, the extension of homology theory to general spaces during the third and fourth decades of the 20th century (Alexandroff, Čech, Lefschetz) was plainly scheduled to occur and created the multiple which, by hindsight, should have been expected to occur.

As Merton (*loc. cit.*) and others have demonstrated, the occurrence of multiples is the *rule,* rather than the exception, not only in mathematics but in science in general. When a cultural system grows to the point where a new concept or method is likely to be invented,[5] then one can predict that not only will it be invented but that more than one of the scientists concerned will independently carry out the invention. Of course, plagiary can occur and has occurred, but it is the exception rather than the rule.

Scientists, and mathematicians in particular, tacitly recognize the fact of multiples in their struggles to achieve priority of publication. It is well recognized that original discoveries, especially if related to an important and rapidly developing area, will usually be made by more than one researcher.

2.   "Clustering of Genius"

Narrowly related to the occurrence of multiples is the so-called "clustering of genius," a phenomenon characterized by sudden outbursts of creative activity, participated in by several researchers whose accomplishments are so remarkable as to warrant applying to them the label "genius." One "explanation" of the phenomenon is that genetic accidents have contributed to simultaneous births of potentially "great men," who subsequently created the phenomenon because of their great powers of intellect.

Unfortunately for this explanation, what is known about genetics today fails to support it. We can refer here to the question from Kroeber already made in connection with our remarks about "great men" in § 3 of

[4]We refer here to such events as the discovery of the space-filling curves, introduction of the notion of local connectedness in general spaces, and Schoenflies' characterization of the continuous curve in the plane.
[5]The common phrase is, "the concept is in the air," meaning, of course, "in the culture."

Chapter I. Even more impressive is the following quotation from the late anthropologist L. A. White:

> In the process of cultural growth, through invention or discovery, the individual is merely the neural medium in which the "culture" (in the bacteriological sense) of ideas grows. Man's brain is merely a catalytic agent, so to speak, in the cultural process. This process cannot exist independently of neural tissue, but the function of man's nervous system is merely to make possible the interaction and re-synthesis of cultural elements. To be sure, individuals differ just as catalytic agents, lightning conductors or other media do. One person, one set of brains, may be a better medium for the growth of mathematical culture than another. One man's nervous system may be a better catalyst for the cultural process than that of another. The mathematical culture process is therefore more likely to select one set of brains than another as its medium of expression.
>
> There must be a juxtaposition of brains with the interactive, synthesizing cultural process. If the cultural elements are lacking, superior brains will be of no avail. There were brains as good as Newton's in England 10,000 years before the birth of Christ, at the time of the Norman conquest, or any other period of English history. Everything we know about fossil man, the pre-history of England, and the neuro-anatomy of *Homo sapiens* will support this statement. There were brains as good as Newton's in aboriginal America or in Darkest Africa. But the calculus was not discovered or invented in these other times and places because the requisite cultural elements were lacking. Contrariwise, when the cultural elements are present, the discovery or invention becomes so inevitable that it takes place independently in two or three nervous systems at once. . . . . . . just as a "brilliant" general is one whose armies are victorious, so a genius, mathematical or otherwise, is a person in whose nervous system an important cultural synthesis takes place; he is the neural locus of an epochal event in culture history (L. A. White, 1947: 298–299).

The flowering of mathematical research during the present century is certainly not due to genetic accident; rather it is due to the multiplication of new fields (especially since World War II), the greater accessibility of learning through the establishment of new universities and libraries, and the greater opportunities for diffusion of knowledge through modern technology.

3. The "before his time" phenomenon

Of special interest in the history of science is the bursting forth, every now and then, of concepts or ideas which fail to attract attention at the time, but which years later are re-created by other researchers, their importance recognized, and their incorporation in the growing relevant field resulting as a matter of course. In genetics, the classical example is Mendel, the significance of whose theories was not recognized until nearly 40 years later (in 1900), when they were re-created by at least three other researchers working independently. An excellent example in mathematics is the history of projective geometry, originally created by Girard Desargues in the early 17th century, but eventually forgotten until its re-creation in the early 19th century by Poncelet and others.[6] In this case, the conception of mathematics as a cultural system, composed of vectors each exerting its own magnitude and direction, is very useful as an explanatory device. At the time, the algebraic and geometric vectors were actively joining to produce analytic geometry. At the same time, the geometric vector, recently strengthened by the projective methods of Renaissance artists, engineers and map-makers, was evidently stimulating its mathematical representatives to apply projective methods in pure geometry. As we shall see in more detail later, this stimulus managed to break through in the person of Girard Desargues, a French architect and military engineer. However, due to forces whose character we do not describe at this time, the efforts exerted by Desargues and his few followers (notably Blaise Pascal) came to nought. The successful breakthrough was, however, finally accomplished, as remarked above, in the early 19th century.

Generally speaking, the "before his time" phenomenon is a natural result of stresses being imposed by a cultural vector V at a time when successful breakthrough of V is smothered by other, stronger vectors; later, at a more opportune and usually more appropriate time, V achieves recognition and consequent development of its proper niche in the evolution of the related science. As will be made clear later (Chapter VI), the initial stress may be due to the accident of there existing an individual who possesses a unique combination of experience and knowledge which forms a perfect receptor for the cultural forces stirring for recognition.

4.   The operation of cultural lag in mathematics.

Cultural lag — the failure of a culture to adopt, or adapt to, innovation — has been studied and discussed by anthropologists and sociologists at great length. It is somewhat analogous to procrastination on the individual level, as well as to conservatism. But while it is usually possible for an individual to overcome procrastination (only *one* individual being

---

[6]Later we consider this case more in detail. See Chapter VI.

involved), on the cultural level it presents a quite different problem. The current attempt to convert the United States culture to the use of the metric system furnishes a good example. There is no good reason for opposing the change in this case, while there are many advantages that would accrue from it. Yet the attempt seems not to be achieving its objective.

As a general phenomenon, cultural lag has what the anthropologist calls "survival value." As Kroeber stated: "A culture that was so unstable and novelty-mongering that it could continually reverse its religion, government, social classification, poverty, food habits, manners and ethics — would scarcely seem attractive to live under ... and it would presumably not survive very long in ... competition with more stable cultures ... " (Kroeber, 1948: 257).

Probably the most elementary examples of cultural lag in mathematics relate to numeral systems — roughly, the way people count. The Greek alphabetic numerals, the so-called "Ionic numerals," furnish a good example. These numerals, consisting of the letters of the Greek alphabet augmented by three archaic letters, with modifying symbols attached such as an accent, were extremely easy to use and were quite sufficient for the ordinary calculations of daily life. Despite the fact that other numeral systems, more capable than the Ionian, such as the place value numerals of Babylonia and, later, the Hindu–Arabic numerals, were known to the Greeks, the Ionian system persisted not only throughout the Greek period but on into the East Roman Empire until the 15th century (Struik, 1948: I, 79). Even the clumsy roman numerals survived long after the demise of the Roman Empire.[7] Often cited is the story of the edict issued in 1299 to the members of the merchant guild in Florence ordering them to cease using Hindu–Arabic numerals and to return to the roman numerals.

Some types of cultural lag are more aptly designated "cultural resistance," especially when the refusal to adopt an innovation is more overt. It has not infrequently occurred in mathematics that despite advantages displayed by new methods, refusal to employ them has been quite aggressive. The long time during which Newton's calculus symbols ("dotage") persisted in England, although the Continent had adopted the operationally more effective Leibnizian notation, seems to have been a case of cultural resistance due to national pride.

---

[7]We do not refer here to their survival, to the present day, for ceremonial and other purposes, but to their every day use.

It can be expected that whenever new concepts are proposed in mathematics that run counter to the prevailing *mores* regarding mathematical existence and reality, then they will be rejected at first. The example of the long-time rejection and avoidance of "imaginary" or complex numbers is fresh in the minds of science historians,[8] as is also the battle over the admission of the cantorian transfinite. The latter was at first even counter to Cantor's own philosophical beliefs, but he was forced to admit it in order to solve a problem in the representation of functions by trigonometric series.[9]

5.   Patterns of thought. Mathematical reality and mathematical existence

It need hardly be remarked that a great deal has been written on these subjects, particularly from a philosophical point of view. Also, it need hardly be remarked that many different conclusions have been reached regarding them. We shall not attempt to give a summary of all the different philosophies that have been proposed, but will content ourselves with two or three examples.

Certainly *platonism* has found the greatest acceptance among mathematicians, amounting almost to a "mathematical theology" in many cases. According to this philosophy, mathematics exists in a world of ideals, ready and waiting to be discovered by the investigator. Numbers, for example, are already existent in this ideal world, even though many (infinitely many) numbers have never been cited in human discourse. Or consider axiomatic systems, in particular a set of axioms for Euclidean geometry. Although a host of theorems have been discovered on the basis of such a system, one can maintain that "there exist theorems of Euclidean geometry that have never been discovered."[10] This type of statement is in accord with the

---

[8]M. J. Crowe (1975) used this and the rejection of incommensurables by the Greeks to illustrate his second law: "Many new mathematical concepts, even though logically acceptable, meet forceful resistance after their appearance and achieve acceptance only after an extended period of time."

[9]Cf. also Crowe's law 1, *loc. cit.*

[10]Although Euclidean geometry is what many mathematicians call "well worked out," there are still mathematicians who specialize in it and who every now and then come up with an interesting theorem that has never been stated before. An interesting case was that of "Morley's Theorem;" see C. A. Oakley and J. C. Baker, 1967. Also see Coxeter and Greitzer, 1967, and Klee, 1979.

philosophy of platonism. Of course, one can avoid such a commitment simply by stating, "The axioms of Euclidean geometry imply more theorems than have been stated to date." It should also be remarked that even those mathematicians who do not profess platonism, or in some cases who even deny the philosophy, do manage to *act* and *talk* like platonists in their daily activities. A mathematician who asks a colleague, "Do you think there exists a function with such and such properties?" is talking like a platonist, even though he may not be one. And one who spends hours and days searching for a counter-example to some proposition is acting like a platonist, in that he is assuming that such an entity may exist.

At first glance seemingly close to platonism, was the philosophy of the famous mathematician Kronecker, if one can judge by his much quoted statement, "The integers were made by God; all else is the work of man."[11] However, Kronecker insisted that only those mathematical entities which are constructible by finite methods on the basis of the natural numbers are admissible. Consequently Kronecker's mathematics led to admissibility of algebraic numbers, but not of arbitrary real numbers. And in the form called "Intuitionism" later developed by Brouwer,[12] mathematical existence came to be associated with a restricted form of constructivism which denied validity to much already existing mathematics. Perhaps the best way to characterize Intuitionism, as well as other types of constructivism, is to call it a "mathematics of doing;" one admits mathematical existence only to those structures — numbers, functions, etc. — that one can conceivably construct.[13]

Whatever form one's conception of mathematical reality might take, it can hardly escape its cultural background. This is, of course, all too evident in the early stages of mathematical history, when mathematics consisted of

---

[11]*Jahresberichte der Deut. Math. Vereinigung,* Vol. 2, 1892. According to E. T. Bell (Bell, 1951: 33), this statement should not be taken too literally, since it was made in an after-dinner talk. In a more satirical vein, perhaps, R. Thom remarked: "This maxim reveals more about his past as a banker who grew rich through monetary speculation than about his philosophical insight."

[12]Expositions of Brouwer's mathematics may be found in A. Dresden, 1924; Wilder, 1952: Chap. X; Heyting, 1956; and in Brouwer's *Collected Works* (Brouwer, 1976).

[13]Actually this does not mean structures that one can build "manually;" a number might be too large to construct this way, for instance. But any given number, no matter how large, is theoretically constructible, even though the best computing machine could not compute it within one's lifetime.

a few numerical devices and elementary geometric formulas for calculating lengths and areas. As V. G. Childe observed (Childe, 1946: 102) of the early Babylonian laws of arithmetic and geometry, "They are too patently products called forth by the needs of a society affected by the urban revolution." The culture and its needs quite clearly determined the form later taken by these early mathematical concepts. It was only when mathematics became an institution in which the practitioner began to invent concepts based on already existing mathematical forms and seemingly independent of external needs, that questions concerning its reality began to arise. Particularly when the notion of infinite and infinitesimal were conceived, did questions concerning existence and reality appear. It was as though one had taken too much to heart the apparent freedom from material matters and ventured too far from practical reality. But in the study of continuous motion, where the above notions seemed to demand recognition, the resultant mathematics seemed to *work*. The symbolic apparatus devised for the calculus, especially by Leibniz, forced the invention of philosophical explanations which could hardly stand up under close inspection. Nevertheless, one continued to use differentials and higher differentials because they gave results in mechanics and geometry that seemed sufficient justification despite their lack of adequate foundation. But under the influence of the Euclidean tradition, mathematics had become a "logical" science, and it seemed illogical that meaningless symbolic modes should give correct results. As a rational being, man wanted to know *why* they worked.

When certain problems in mathematical analysis compelled 19th century mathematicians such as Cantor and Dedekind to invent concepts of the actual infinite for their solutions (see § 7 below), and especially when the too much extended exploitation of the resultant concepts led to logical contradictions (Burali–Forti, Russell), then indeed did the stress for justification of their "reality" and/or existence become too strong to ignore. The result has been the introduction of a number of competing philosophies concerning just what parts of mathematics are "real" and, more to the point, can be developed within the bounds of consistency. Presumably, one would say, on the basis of platonism, that this constitutes a search for the "true" mathematics, and quite likely a consensus among mathematicians will be reached in the future regarding the matter. However, partly on the basis of present tendencies, and partly on the basis of the increasing growth of computers and computer theory in both mathematics and its applications, this consensus may take the form of a constructive mathematics. On the other hand, the recent developments in non-standard analysis, providing a justification for the modern forms of Leibniz's theory, as well as powerful and simplified methods of proof in many cases, can lead one to predict their ultimate acceptance.[14]

Whatever the future may bring regarding these questions, there would seem no denying that what we are dealing with is cultural in nature. Some express this by saying that "mathematics is man-made." There seems little justification for not including mathematics among the other forms of cultural elements — political institutions, religions, customs and the like — for mathematics is certainly one of the cultural devices that man has created both for adaptive purposes and for his own intellectual satisfaction.[15] The advantages in taking a culturological view of mathematics are to be found not only in the perspective it affords for the understanding of the various philosophies of mathematics and their *raison d'être*[16] but in the consequent disclosure of the cultural forces that have dominated and still control the evolution of mathematical invention.

6. Evolution of greater abstraction

As a cultural system becomes institutionalized and achieves greater growth, there inevitably results greater abstraction. This can be considered a cultural law and is not restricted to scientific systems. For instance, religious systems whose initial stimuli were "prophetic," adopt simple rules and rituals which, as the religion grows, become augmented by more abstract "theologies" in order to meet all the exigencies of complex societies. Even political systems, as, for example, that of the United States, grow more complex and abstract with advancing time. Similar remarks hold for the various philosophies that have been proposed, as well as for legal systems.

The growth of modern science, including mathematics, has been

---

[14]Just how much effect non-standard concepts will have in other parts of mathematics, such as topology and modern algebra, is difficult to predict at the present.

[15]A most convincing and scientific exposition of the cultural nature of mathematics may be found in the article White, 1947.

[16]See Wilder, 1950 for further commentary.

characterized by the same tendencies to greater abstraction. Every science, by definition, develops *theories;* theories constitute the heart and soul of science. Even those sciences that commenced as collecting and classification systems have ultimately developed theories explanatory of their findings. As I have previously remarked, it is characteristic of science that "the farther its concepts seem to recede from external reality, the more successful do they become in the control of man's environment" (Wilder, 1959). Greater abstraction leads to greater power. [17]

It has been my fortune to observe, practically from its beginnings, [18] the growth of one of the most important fields of mathematics, viz. topology. The increasing abstraction (which I at first deplored, in my ignorance of evolutionary tendencies), which its growth early exhibited, was continually a source of amazement to me. It became quite apparent to me, as the field developed, that the more abstract it became, the more powerful were the tools that it developed. Also, with increasing abstraction, a broader perspective was achieved, and earlier results that occupied special niches in the field became a natural part of a higher, unified theory.

These phenomena, as anyone familiar with the history of mathematics should recognize, were not peculiar to topology, but were characteristic of other fields, of which modern algebra is an excellent example. The latter has grown to such an extent that its abstract concepts find application in all branches of mathematics, including topology. It is surprising, however, how many mathematicians, even historians, have deplored increasing abstraction. For example, Coolidge, in his excellent book *A History of Geometrical Methods* (Coolidge, 1963: 251), after sketching such topics as the Menger–Urysohn dimension theory and von Neumann's continuous geometry, states:

> It is evident that in all of this we are skating around the outskirts of abstract geometry, abstract algebra, point-set theory, large disciplines that have been pushed far in the second and succeeding decades of the twentieth century. An extended

---

[17]Compare A. N. Whitehead's remark: "The paradox is now fully established that the utmost abstractions are the true weapons with which to control our thought of concrete fact." (Whitehead, 1933; or 1948: 34).

[18]Although the earliest results that can be termed "topological" were found during the second half of the nineteenth century, the subject did not become a recognized discipline until the second and third decades of the present century.

discussion of abstract geometry will be found in Moore R. L. (reference here to Moore, 1932). The whole joins the general theory of topology .... It is widening to the horizon to see that a doctrine which at least uses some of the language of geometry can be built up in a fashion that is independent of coordinates, of measurement, of the number system almost. Yet I must confess that there remain in my mind certain old-fashioned doubts. We drift in this fashion far indeed from anything suggestive of space intuition.[19] Moreover, a certain class of mathematicians, to which I personally belong, finds it hard to maintain interest in any mathematical theory which does not seem to connect with any concrete problem, or anything related to our world of sensible objects. If lines in S_4 mean circles in our space, if points of a V_4^2 mean our lines, that conveys something natural to the mind. But a theory of abstract spaces that covers 300^{20} pages! The dimensions that really count can be bought cheaper.

We should remark, at this point, that those who fear that the abstractions of modern mathematics are carrying the field too far away from "applications," should be reminded that not only have its abstractions inevitably led to greater power in a mathematical sense, but have gradually become a fertile source for concepts useful to the more advanced sister sciences, especially physics (cf. Wigner, 1960). I have frequently cited von Neumann's statement that "by and large it is uniformly true in mathematics that there is a time lapse between a mathematical discovery and the moment when it is useful; and that this lapse of time can be anything from 30 to 100 years, in some cases even more; and that the whole system seems to function without any direction, without any reference to usefulness, and without any desire to do the things which are useful" (von Neumann, 1961). Not only have some of the most abstract theories been found "useful," but (as in the case of matrix theory and quantum mechanics), have anticipated applications in which their substance has later been worked out independently by physicists.

It is the increasing abstraction of a special field which makes possible its permeation of other fields. This has been the case with topology, for example, which started as a special field of geometry, but which, in its modern abstract forms, has been adopted by other fields of mathematics as a means of expanding their own theories. The ultimate result of such developments is the greater unification of modern mathematics.

[19] A set-theoretic topologist would hardly agree with this. R. L. W.
[20] If Collidge is referring here to Moore's book, it contained 486 pages.

7. Forced origins of new concepts

It is noteworthy in the history of mathematics that in order to save useful symbols or ideas which for various reasons seemed inadmissible, the creation of new concepts has been compelled in order to overcome their objectionable features. This process we shall call *conceptual stress.* The most elementary and best known example is probably the symbol $\sqrt{-1}$. Its occurrence in the solution of third-degree algebraic equations seems to have been the initial stress created by this "unreal" symbol. Additional considerations, such as the apparent fact that only by admitting it to number status could one assert that every algebraic equation of degree n, $n \geqslant 1$, has precisely n roots, ultimately forced the invention of acceptable definitions of complex numbers involving $\sqrt{-1}$. First, geometric interpretations of these numbers were given — by Caspar Wessel in 1797, by Jean Robert Argand in 1806, and by Gauss. Gauss's representation was published in 1831, although he had earlier, in 1811, described it in a letter to Bessel. Algebraic interpretations soon followed; W. Rowan Hamilton in 1837 defined complex numbers $a + bi$ as ordered pairs (a, b) of real numbers a and b, along with rules for operating with such pairs.

There is also the classic example of Eodoxus's theory of proportion, apparently conceived in order to allow comparison of arbitrary geometric magnitudes. In this case it was the Greek discovery that not all such magnitudes are expressible as ratios of integral magnitudes that formed the initial stress. The example is of special interest because of its antiquity. As one approaches the modern period, the cases become more numerous.

An outstanding modern example is afforded by the introduction of the completed infinity by G. Cantor during the last quarter of the 19th century. Until this time, the notion of a completed infinity had been generally rejected. As Gauss stated, the infinite was "merely a way of speaking." Absurdity seemed to result from admitting even the complete infinity of natural numbers, as Galileo discovered (1638); namely the "numerical equality" of the collection of all the natural (counting) numbers and the collection of their squares; that is, to each natural number n one can make correspond its square n^2, so that "counting" the series $1, 4, 9, 16, \ldots$ uses up all the natural numbers. Of course, one must assume the completed totality of all natural numbers for this to hold. (Cf. EMC: 118, EMC_1: 115.)

A geometric case, equally "objectionable," consists in the pairing of the points of a straight line segment — say of length 1 — with the points in the half-interval. This can be done by constructing an equilateral triangle *abc* on the base *bc* of length unity, and a segment *de* connecting the midpoints of *ab* and *ac*, respectively. For each point *p* of *bc* the line *ap* intersects *de* in a point *p'* and the parallel through *p'* to *ab* interects the base in a point *p''*. Then every point *p* of *bc* finds in this way a unique correspondent *p''* in the half-interval *bf*.

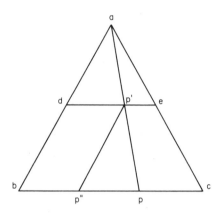

Cantor's achievement was to introduce the notion of *number* for infinite sets in such a way that these seeming "absurdities" turn out to be quite natural and acceptable.[21] This concept was a part of his general theory of sets which, although successful in solving fundamental problems in mathematical analysis, was at first met with severe cultural resistance. Its inevitable acceptance and admission to mathematical respectability was a result of its power in extending modern analysis and topology, whose development would have been impossible without the Cantorian theory and its extension by later researches. However, as pointed out in § 5 above, there has been renewed activity on the part of so-called "constructivists" to modify the "classical" theory of sets; but this is a phenomenon quite apart

---

[21]Actually, Cantor first introduced the completed infinite as a tool for solving a certain problem concerning the representation of functions by infinite series. In other words, he was not playing an interesting game, but creating a tool for solving a mathematical problem that could not be solved otherwise.

from the exemplification of the operation of conceptual stress in the formation of Cantor's ideas.

The "moral" of such examples as those just cited seems to be that whenever the advance of mathematical evolution requires the introduction of seemingly absurd or "unreal" concepts, the latter will be provided by the creation of appropriate and acceptable interpretations. Regarding the mechanism of concept formation in mathematics, see my comments in § 4a of Wilder, 1953.

8. Selection in mathematics

"Selection," in the form of "natural selection," has become well known through popular discussions of Darwinism and of botanical phenomena, but it should be pointed out that we are not introducing the notion here in an attempt to "ape" the successful use of the notion in the natural sciences. For the type of selecting which occurs in mathematics may or may not be analogous to natural selection, in that in most cases it has special features.

We mentioned above (§ 5) the present situation in mathematics where in addition to the classical type of mathematics there have been proposed radically different types called "constructive" and "non-standard." While currently the majority of mathematicians, in both their teaching and research, employ classical methods and theories, it is quite conceivable that future developments will lead to a selection of either the constructive or non-standard approaches to mathematical foundations, or even some as yet not proposed alternative to the classical type of mathematics. This would seemingly constitute a kind of selection analogous to the Darwinian type.

Just how cultural systems make selections has not, so far as I know, been studied by anthropologists. In the Darwinian theory, selection is characterized by the so-called "survival of the fittest." While such a criterion might explain some cases of selection in mathematics, in many other cases it would be difficult to classify it in this manner. For example, a mathematical theory emanating from a so-called "important" mathematical center has usually had greater chance of mathematical survival than one proposed by a lone worker. The case of Grassmann and Wm. Rowan Hamilton is illustrative of this: Hamilton discovered quarternions in 1843; Grassmann published his so-called "Ausdehnungslehre" in 1844, containing similar concepts of non-commutative algebras. However, Hamilton was not only a member of Trinity College, Dublin, but had earlier distinguished himself by his research on crystallography, resulting in his being knighted at the age of 30; but Grassmann was a relatively obscure teacher in a German secondary school. While Hamilton's work was immediately reported to the Royal Irish Academy and received instant recognition, Grassmann's work was only slowly recognized. It is true, too, that Grassman's work was couched in an unconventional terminology, making it difficult to read; clearly if one has to choose between different expositions of a theory, one of which is presented clearly and in language of a conventional type and the other necessitating adapting to a new langauge, the former is the more likely to be selected.

In the case of ideographic symbols and terminology, the value of being at an important mathematical center where one has colleagues who can be immediately apprised of the new terminology is widely recognized. In such cases it is hardly possible to term the selection as a case of "survival of the fittest."

The reasons for selections differ from one case to another. While the selection of a general theory may at first be chiefly influenced by the eminence of the author and the status of the institution with which he is connected (both cultural features), its survival in the long run is more dependent upon its mathematical significance; i.e. its usefulness for other theories and especially its ability to further the general evolution of mathematics. In this respect, then, that is in the survival of general theories, selection in mathematics is quite similar to the operation of natural selection in biology. But it would be difficult to attribute the survival of such ideograms as π, e, i (for $\sqrt{-1}$) to their "fitness" in contrast to other possible symbols. Here it was the eminence of Euler and the importance of his work that influenced their adoption (cf. Boyer, 1968: 484, for example). The term "analysis" for the study of infinite processes, preserved today in the name of one of the major fields of mathematics, was also evidently due to the use of the term in Euler's *Introduction in analysis infinitorium.*

When an author of a theory needs new symbols for the various constants and variables involved, he presumably chooses symbols which he has become accustomed to using in his development of the theory. If the theory itself survives, then the symbols so chosen will ordinarily continue to be used. Here individual choice appears to govern. On the other hand, designation of points in Euclidean geometry in English treatises has usually been by capital roman letters A, B, C, etc., and of lines by lower-case letters such as l_1, l_2, etc. But individuals familiar with classical logic, and especially theory of types, usually reverse the process, using lower-case letters a, b, c, ... for points and capital letters L_1, L_2, etc., for lines (and for collections of lines capitals of another type, as e.g. German letters) clearly a cultural development.

In short, selection on a "global" basis, as in the case of general theories which have evolved at the hands of several workers over an extended period of time, has usually been a cultural process, where as for symbolic purposes, both individual choice and cultural selection have governed the selection.

9. The effect of the occurrence of paradox, or the discovery of inconsistency

It has quite uniformly been the case in mathematics that the discovery of paradox or contradiction has resulted in the invention of new theories to remedy the situation.

Traditionally, mathematics cannot tolerate contradiction.[22] If it is discovered that a concept or theory leads to contradiction, then either modification of the offending concept or theory will ensue, or it will be abandoned. In the use of the method of *reductio ad absurdum,* if the negation of a proposition can be shown to lead to contradiction, then the proposition is considered as "proved."[23]

Sometimes contradictions are only apparent, having been derived from hidden assumptions. The classical case is that of the previously cited example of the (1–1)–correspondence between the natural numbers and their squares. This and similar examples led to the inference that the concept of the completed infinite leads to contradiction. However, this inference was based on the assumption that since a finite collection cannot be put in (1–1)–correspondence with a proper part of itself, then the same must hold for infinite sets. It was Dedekind who recognized that this so-called "contradition" was only a paradoxical property of infinite sets;

---

[22]This is, of course, the basis for the famous method of proof known as *contradictio ad absurdum.*

[23]Modification — and in some cases outright rejection — of this type of proof occurs in some of the constructive philosophies of mathematics, especially Intuitionism; see II-5.

moreover, that it could be used as a characterizing or defining property of infinite sets, a set S being infinite if and only if it has a proper part S_1 such that between the elements of S and those of S_1 a (1–1)–correspondence can be established (or "exists"). We have already described, in § 7, the subsequent development, at the hands of Cantor, of a theory of sets embodying the notion of the completed infinite.[24]

It later turned out, to be sure, that the theory of sets did harbor contradiction (see Wilder, 1965: 54–58), but this did not result in abandonment of the theory — although had it not proved so useful in solving problems and engendering new concepts, it would most certainly have been abandoned. Rather, ways were found to avoid contradiction by suitably limiting the theory; that is, roughly speaking, by allowing only infinite sets which are known not to exceed certain limits of size,[25] contradictions of the sort discovered no longer appeared.[26]

The situation is quite analogous to that commonly found in engineering. If an invention does not work — i.e. carry out the tasks for which it was devised — then modifications are sought to enable it to achieve its purpose. The history of the heavier-than-air flying machines is an excellent example. Accordingly, it can be predicted that whenever a new mathematical theory is found to lead to contradiction, or for any other reason fails to achieve its objectives, then — assuming the need for it is great enough — it will be so modified as ultimately to accomplish its purpose.

Naturally, as implied above, if the need for a theory is not sufficient and it is found to harbor contradiction, then it will inevitably be abandoned — especially if the contradiction is not correctable. This seems not unlikely to occur when one experiments with alterations of existing theories which appear to have attractive possibilities. Indeed, it has happened in

[24]For a historical footnote regarding the influence of Dedekind on Cantor, see Grattan-Guinness, 1971.

[25]Note for the non-technician: That infinite sets occur in his theory in different sizes was one of the basic discoveries of Cantor. In particular, the infinity of points on a straight-line segment is greater, in a well-defined sense, than the infinity of the natural numbers 1, 2, 3, ...

[26]Of course, it might happen that new contradictions would occur, previously undiscovered, in which case further limitations would presumably be imposed on the theory of sets. However, the current theory of the infinite, as embedded in well-known systems of axioms (such as the "Zermelo–Fraenkel axioms," or the "von Neumann–Bernays axioms"), have successfully withstood the test of time so well, that mathematicians are generally confident that this will not occur. Rejection of the theory of the infinite on philosophical grounds is, of course, another matter entirely.

mathematics, that theories have been so conceived, and developed to sufficient extent to warrant publication, only to be later found to contain contradiction. In such cases, the need not having been established, the theory is simply abandoned.[27]

The fact that occurrence of paradox or inconsistency in a theory, will, when the need for the theory is great enough, serve as a spur for research, either to find a suitable modification of the theory, or to find an alternative theory that will serve the same purpose, needs to be emphasized. A classical case appears to have been the response of Greek geometry under the stresses caused by the discovery of irrationals and the Zeno paradoxes. In the already cited case of the theory of sets (§ 7), the response was a series of proposed new foundations for mathematics itself in the forms of Intuitionism, Formalism and Logicism, as well as (the successful) exploitation of the axiomatic method. During the first third of the 20th century, while the various mathematical specialties continued to carry on research in the traditional manner, there existed generally, throughout the mathematical community, an interest in the developing "schools" of thought regarding the possible attainment of a contradiction-free mathematics including the new theory of sets. Even the great John von Neumann, whose achievements in the fields of computing and nuclear armaments have been well publicized, participated in the axiomatic developments in the theory of sets (as witness the "von Neumann–Bernays theory," already mentioned above.)[28]

10. The relativity of mathematical rigor

It seems to be a commonly held belief, chiefly outside the mathematical community, that in the realm of mathematics can be found absolute truth. Such phrases as "mathematical truth," "mathematical certainty," are to be found within mathematical circles, to be sure, but usually with generally understood qualifications. Now it is certainly the case that according to the

---

[27]A relatively recent case of this sort was Yoneyama, 1917–20. In this paper he studied indecomposable continua having only finitely many composants. However, Janiszewski-Kuratowski, 1920 proved that every indecomposable continuum has infinitely many composants.

[28]For a general discussion of the rise of "schools" of foundations, the reader is referred to Kline, 1972: Chap. 51.

ordinary meaning of "truth," as used in the market-place, such formulas as "2 + 2 = 4" are true — i.e. in the sense that if to 2 apples in a box are added 2 more apples, then there will be 4 apples in the box. That is, within certain contexts — in this case, applications of mathematics — such formulae are true. But there are differences encountered when mathematical formulae are confined to mathematical theory, rather than to applications.

A striking example of this kind was worked out recently by the late mathematician Imre Lakatos (Lakatos, 1976). He took the Eulerian formula "$v - e + f = 2$," where v stands for the number of vertices of a polyhedron, e the number of edges and f the number of its faces, and, in a remarkable series of essays, gave successive "proofs" of this formula, only to show in each case that there are exceptions to it. In brief, despite the addition of more specific instructions regarding the kinds of polyhedra that were intended, and "proofs" thereof, new types of polyhedra were turned up for which the formula is false. Yet in each case the "proof" appeared to be quite rigorous at the time it was described.

In view of such demonstrations, as well as of other well-known historical cases, it becomes apparent that "proof" in mathematics is a culturally determined, relative matter. What constitutes proof for one generation, fails to meet the standards of the next or some later generation. Yet the mathematical culture of each generation possesses generally accepted standards for proof. At any given time, there exist *cultural norms* for what constitutes an acceptable proof in mathematics. And what is acceptable in one period may not be acceptable at a later period.

It is ordinarily asserted that mathematical proof rests on *logic.* Ideally, this may be correct; actually it is generally incorrect. Analyze the general run of mathematical proof and it will be found to contain hidden assumptions of a mathematical character; these assumptions are generally accepted in the contemporary mathematical culture. The classic example is that of Euclid, whose geometry was for centuries held up as the ideal example of rigorous proof. We now know that it contains geometric assumptions, unstated, that invalidate some of the proofs, and, in some cases, render the theorems *false.*[29] Even the proof of the first theorem is inadequate, as many writers have remarked.[30] As mathematics grows, such

[29]Proposition 16 of Book I. See Dover ed. of Heath, 1956: Vol. 1, pp. 279–280.
[30]See, for example, Russell, 1937: 405–407.

hidden assumptions are unearthed and made explicit, resulting in general acceptance or rejection; acceptance usually follows analysis of the assumption and justification by accepted methods of proof.

An example is the long-accepted rule that a continuous real function which is negative at a real number a and positive at a number $b > a$ must be zero at some number between a and b. The geometric equivalent is the assumption that a line which "crosses" another line must have a point in common with the latter — an assumption which analysis depended upon for a long time as justification for the former. However, 19th-century analysis came to recognize (Bolzano, Cauchy) that the continuity of line and of real number system requires justification (a process in which most outstanding mathematicians of the century participated). Each generation of mathematicians finds it necessary to justify (or reject) the hidden assumptions of the preceding generations.[31]

11.   Growth patterns of fields of mathematics

During its early history, mathematics can be roughly divided into the fields of arithmetic (including number theory), geometry and algebra.[32] The emergence of trigonometry, in the Hellenistic period, presaged the formation of a new field, viz. analysis. With the decline of the Greek culture, a virtual hiatus ensued in the development of mathematics — due to what we term *environmental stress.* As we pointed out earlier, this hiatus was not due to any dearth of mathematically capable minds. As a part of the general culture, mathematics inevitably declines whenever the culture in which it is embedded does.

With the 17-century work of such mathematicians as Newton and Leibniz, however, there was a noteworthy resumption of mathematical creativity, especially in the field of analysis, which continued throughout the 18th and 19th centuries. With the beginning of the 19th century, there was a resurgence in geometry partly with the discovery by Poncelet and

---

[31]For a general discussion of the relativity of mathematical rigor, see Wilder, 1973: Vol. 3, pp. 170–177.

[32]This is from the modern standpoint, of course. Different cultures and time periods have included material in "mathematics" that we would not today include, as, for instance, music. The earliest extant history of mathematics, that of J. E. Montucla, 1758, included much of what we today call applied mathematics or engineering.

others of projective geometry (see Chapter IV), and partly with the consolidation of geometry and analysis to form differential geometry as well as of geometry and algebra to form algebraic geometry. With the introduction of these new fields of mathematics, the evolution of modern mathematics was under way. By the beginning of the 20th century, the seeds of new specialties had been sown. Today such new specialities as geometric and algebraic topology, group theory, functional analysis, statistics, probability, etc., have evolved.[33]

The above may seem to be a picture of general, continuous growth, except for the sparse interval between the Hellenistic and Renaissance periods. But this is due to the brevity of the outline as we have given it. The anthropologist A. L. Kroeber made an extensive study of the growth patterns of scholarly systems, including philosophy, science (which included mathematics), philology, etc. (Kroeber, 1944) He was unable to find any distinctive common growth patterns in these fields. A study of mathematical history would probably make similar findings in mathematics. Each field of mathematics seems to establish its own pattern of growth. However, although there is not a growth pattern common to all mathematical fields, there are growth features that appear to dominate the historical picture.

Consider Euclidean geometry, for instance. From an early smattering of results consisting chiefly of formulae for lengths and areas, created to meet the practical needs of farmers, astronomers, etc., geometry underwent a period of growth until it met a crisis (discovery of incommensurables, uncovering of hidden assumptions), and then matured through the introduction of axioms as a basis for a systematic logical derivation of the system as exemplified in Euclid's *Elements*. Ultimately reaching a state where the discovery of new theorems became difficult, geometry became dormant except for scattered results obtained by the Arabs. It left a legacy of unsolved problems, however, such as the "squaring of the circle," "angle trisection," and "duplication of the cube" (Delian problem), as well as the question of the status of the parallel axiom (Euclid's 5th Postulate). In the post-Hellenic period, much frustrating work was devoted to these problems, resulting in little apparent significant advance in the field of geometry.

[33]See, for instance, the listing of specialties in the journal *Mathematical Reviews,* which publishes abstracts of papers in the various fields of mathematics.

During the early 17th century, geometry enjoyed a revival, chiefly through the consolidation with algebra to form analytic geometry (Descartes, Fermat), a subject which proved useful to the analysts of the succeeding generations. As already remarked, geometry achieved a tremendous boost in the 19th century, through the introduction of projective methods and consolidations with algebra (algebraic geometry) and analysis (differential geometry), as well as in the introduction of the non-Euclidean geometries (the legacy of the problem concerning the status of the parallel axiom of Euclidean geometry).

The general pattern of growth of geometry, then, was birth in practical needs and astronomy, a critical period of self-examination, then development as a subject of study sufficient in itself until an apparently "worked out" status was reached. After a long hiatus chiefly due to environmental conditions, it renewed activity in new forms resulting partly from consolidations with other fields of mathematics, and partly through the discovery of more extensive geometries (projective geometry, non-Euclidean geometries) of which the classical geometry of the *Elements* formed a special case.

The modern theory of sets presents some analogies, in its development, to the manner in which geometry developed. Instead of originating in needs external to mathematics, however, it was born in the 19th-century study of fundamental questions of real analysis, especially through the research of G. Cantor and Dedekind. About the turn of the 20th century it ran into a "crisis" in the form of the set-theoretic contradictions (Burali-Forti, Russell).[34] Ignoring the radical attempts to avoid these by new philosophies of mathematics (Intuitionism, logicism, formalism, newer forms of constructivism), the most generally accepted way out seems to have been sought in the use of the axiomatic method (Zermelo–Fraenkel, von Neumann–Bernays). This has led to a fruitful theory, not so generally accepted as the *Elements,* to be sure, but intensely revealing and productive in contributions to the foundations of mathematics. Moreover, aside from the fulfillment of general mathematical needs, it has already found useful consolidations, such as in the field of set-theoretic topology.

But the occurrence of "crisis," so called, cannot be considered as a typical pattern in the development of mathematical fields. The field of

[34]Cf. Wilder, 1965: §§ 1.1 and 2 of Chap. III, Problem 28 of Chap. IV, and § 3.f.3 of Chap. V.

topology, for instance, while originating in scattered results found chiefly in the 19th century, and in its set-theoretic forms variously axiomatized (Fréchet, Riesz, Hausdorff, R. L. Moore, Kuratowski, ...), seems so far to have escaped such foundations crises as beset early Greek geometry and modern set theory. It has, however, like the two former fields, formed fruitful consolidations with both algebra and analysis (algebraic topology, differential topology).

So far as their origins are concerned arithmetic and trigonometry, the progenitors of classical analysis, followed a similar pattern in that they were created to meet the needs of the "man in the street," and the astronomer. The Greeks separated arithmetic into the "practical" aspects — the arithmetic of the market-place — and what is today called "number theory" (as presented in Books VII–IX of Euclid's *Elements*). The latter is today preserved in modern number theory which, while still concerned with the properties of the natural numbers, has adopted tools from the other branches of mathematics, especially analysis and geometry, through a process of consolidation. The aspects of arithmetic as it is commonly taught in modern schools are properly a part of analysis, although the treatise of Diophantus (the *Arithmetica*) can be taken to represent a beginning of generalization into what we could term "school algebra," especially in a symbolic sense. The extremely influential (in Western Europe) work of al-Khowarismi continued the evolution into algebra, although symbolically not on a level with Diophantus's work.[35]

With the diffusion of Arabic works into Western Europe (chiefly through Spain and Italy), mathematics became a "melting pot" of arithmetic, algebra and geometry. It was a time of improvement of symbolism, especially in algebra — although the final success of the decimal system, long anticipated, occurred during this period. And, especially at the hands of Vièta, trigonometry began to assume a character somewhat prophetic of the theory of real functions.

Geometry received the first impetus toward modern mathematics in the consolidation with algebra by Fermat and Descartes in the 17th century. From this point on, mathematics evolved not according to any discernible pattern, but chiefly through the impact of both external and internal forces.

[35]Al-Khowarismi's work was versed in what modernists call "rhetorical algebra," and was chiefly influenced by the Hindus and Babylonians. Indeed, there seems to be no evidence that al-Khowarismi knew of Diophantus's work.

It seems, then, that although no general patterns of growth are evident, we can detect certain special patterns in the origins of mathematics and its special fields: (1) origins in the other fields of mathematics and/or of sister sciences such as astronomy; (2) consolidation with or absorption by other fields of mathematics, particularly when advanced or "worked-out" status is achieved. We may conclude, although the above remarks are certainly far from exhaustive, that in the internal interplay of mathematical concepts, there seems to be more fruitfulness in the study of particular forces, internal to mathematics as well as external, which affect the development of mathematics, than in the search for a common pattern in the growth of mathematical specialities. This was done in considerable extent in EMC.

12. A problem

An epistemological problem which has puzzled many modern physicists and mathematicians concerns a pattern that the Nobel prize winner Eugene P. Wigner calls "the unreasonable effectiveness of mathematics in the natural sciences" (Wigner, 1960). Stated succinctly in Wigner's words, "the enormous usefulness of mathematics in the natural sciences is something bordering on the mysterious and ... there is no rational explanation for it." He cites in particular the use of complex numbers in the laws of quantum mechanics, concluding "it is difficult to avoid the impression that a miracle confronts us here."

There seems little doubt that the solution to this problem ... the effectiveness of mathematics in the natural sciences ... is to be found in the common cultural origins of mathematics and the natural sciences. Stated from a converse point of view, imagine two cultures, one, C_1, the scientific culture which has been created on this earth, and another, C_2, created by inhabitants of another planet in a portion of the universe where a totally different kind of mathematics[36] had developed. Then quite conceivably the mathematics of C_2 would not exhibit at all the usefulness in the natural sciences of C_1 exhibited by the mathematics of C_1. It is unquestionably due

---

[36]This is purely hypothetical, of course. What such a mathematics might be like is hard to imagine; the geometry, for instance, might be developed along topological lines instead of the metric lines of Euclidean geometry as we know it. In any case, such a mathematics would presumably be identifiable as such through its foundation on the natural numbers, which are generally admitted to be a cultural universal.

to their common origin in our own culture, C_1, that mathematics finds such significance for the natural sciences. When one considers that mathematics was created not only with an eye on the solution of astronomical and physical problems, but that in addition the creators were, during early days, themselves astronomers and physicists,[37] some of the mysteries concerning their present-day intimate connection disappears. Furthermore, throughout its modern evolution, physics has been developed on a mathematical basis. When a mathematical physicist is at a loss for a new conceptual approach, he usually looks to mathematics for a clue; the relation between mathematics and physics has been a reciprocal one, each having afforded ideas for the other. It is predictable, on this basis, that some of the unexpected uses of mathematics in physics would appear to border on the mysterious. Vestiges of many of the intimate associations of ancient cultures, no longer so evident in their modern counterparts, are nevertheless preserved in hidden forms.

[37]Note that Plato's famous pupil, Eudoxus, is almost invariably called a "mathematician" in histories of mathematics, and an "astronomer" in histories of astronomy.

CHAPTER III

Historical Episodes; a Laboratory for the Study of Cultural Change

> We can not help feeling that certain mathematical structures which have evolved through the combined efforts of the mathematical community bear the stamp of a necessity not affected by the accidents of their historical birth.
>
> H. Weyl

The preceding two chapters have contained historical materials and especially historical episodes, as a means for illustrating and/or justifying theoretical assertions. Thus, in discussing the relationship between the individual and his culture in § 3 of Chapter I, experiences of several inventors were cited, and similarly in § 1 of Chapter II some cases of multiple invention were cited. The well-known case of the Lobachewski–Bolyai–Gauss introduction of non-Euclidean geometry was cited both as a multiple and as a case of the inevitable occurring when mathematics was prepared for it, as was also the case of Desargues and projective geometry in II-3 as an instance of the "before-his-time" phenomenon. Cantor's studies of the infinite were mentioned in II-7 as exemplifying the forced origins of new concepts.

In the present chapter the historical is made of prime importance, being treated as a laboratory for the illustration and/or detection of these cultural patterns and forces that have been operative in cultural change. In the brief discussion of mathematics as a cultural system (I-4), the idea of cultural forces was first explicitly mentioned, especially in the form of internal stresses due to conceptual needs. In addition, diffusion and cultural lag as operative in cultural change were mentioned. The chief purpose of this chapter will be to select historical episodes that serve to bring out such stresses.

47

1. The great diffusions

Students of the evolution of culture have been forced to recognize the prominent part that diffusion — the passing of cultural elements from one culture to another — has played in the evolution process. Probably the principal factor in the discovery that cultures do not follow uniform patterns in their evolution was diffusion. One might try to set down the stages through which a culture passes in the development of number systems and arithmetic, for example, only to discover that a specific culture did not pass through all such stages because diffusion of higher stages into the culture allowed bypassing intermediate stages. Especially where missionaries or traders have intruded into a culture, one finds leaps from primitive types of counting to the use of the so-called Hindu–Arabic system with its decimal base and use of place-value.

On the other hand, one can set down stages through which counting systems *as such* evolve, even though a particular culture may not pass through all these stages (see EMC: 180, EMC$_1$: 174).

But to return to the subject of diffusion. The greatest of the early diffusions is indicated by the following diagram:

Babylonia ———→ Greece ———→ Arabia ———→ Western Europe
Egypt   India

Both arithmetic and geometry, as well as elementary algebra (in a primitive form), are involved in this schema. Mathematical formulae used by the Babylonians and the Egyptians filtered through trade and intellectual channels to Greece and India, from whence they diffused to Arabia and through the latter to Italy and Spain. The term "Hindu–Arabic" originated from the diffusion of Hindu numerals to Arabia. Of course, if we were writing a history, we would describe mathematical achievements of the Chinese and Mayan cultures, but so far as we know, these did not influence (absence of diffusion) the development of mathematics in the West. Naturally each culture leaves its own stamp on the elements diffused.

But this, what might be called *geographic* diffusion, albeit typical of anthropological examples is no more important than diffusion between *fields* of mathematics, as well as between mathematics and the natural sciences. An excellent example of this type of diffusion is afforded by the history of logic.

"Discovered" by the Greek philosophers, logic penetrated Greek mathematics quite early through its employment of the axiomatic method. Euclid's *Elements* are held up as the prime example of so-called "logical deduction." In both disciplines philosophy and mathematics, logic passed through its medieval phases until at the hands of such scholars as De Morgan and Boole, it developed into the highly symbolic *mathematical logic,* which in the 20th century was to achieve the status of a mathematical field in its own right. This mathematical logic has diffused widely in mathematics, both into the foundations of mathematics and the mathematics of computation.

This brings us to what is perhaps the greatest diffusion of all, namely, from mathematics to the natural sciences and technology. Without this diffusion of mathematical methods and concepts, our modern technological culture would not exist. Of course, it is a "two-way street;" not only do mathematical concepts infuse natural sciences, but these disciplines in turn suggest models for mathematical theories. These are well-known processes, and only need be mentioned here.

The reasons for, and mechanisms of, special cases of diffusion offer an interesting study, which will not be undertaken here. Although diffusion may occur under stress, as in the well-known cases of customs and religious concepts imposed by military conquest, a common pattern in modern times is that of diffusion to fulfill a need — almost exclusively the case when diffusion occurs within and between scholarly disciplines. Within mathematics, for instance, analysis borrows what it needs from topology; so much of basic topology has permeated analysis that a course in topology is now usually required of the aspiring analyst.

2. Symbolic achievements

In section I-3, we stressed the symbolic basis of communication, the "cement" which binds a culture together and makes possible its evolution. In the course of its evolution, mathematics has become highly symbolized. To study mathematics, the child begins with the symbols 1, 2, 3, ..., which have now become nearly universal. The evolution of these symbols, and of the arithmetic by which we operate with them, is itself a fascinating chapter of cultural evolution.[1]

[1]For details, the reader is recommended to: Conant, 1896 (unfortunately not available in the average city library); Tylor, 1958: VII; Dantzig, 1954; Menninger, 1954 (authoritative); and, for African numerals, Zaslavsky, 1973.

Although every folk culture seems to have yielded to the necessity of counting, symbols were usually created in various forms of tallies, or "finger-and-toe counting;" or, in more advanced cultures, number-words. In more complex societies, where records were needed for taxes, land ownership, building trades, commerce, etc., more efficient modes of counting were necessarily developed. Three great ahievements marked the development of counting ("natural") numbers: (1) cipherization, (2) concept of place value, and (3) invention of a zero. By "cipherization" (a term due to C. B. Boyer) is meant the invention of efficient symbols for individual digits; our Hindu–Arabic digits 1, 2, 3, ..., represent the peak of cipherization in our Western culture. By place value is meant the assignment of a "value" to a digit according to its position; thus, the "values" of "2" in the successive numbers 251, 25.1, 2.51 and .251 are 200, 20, 2 and 2 tenths, respectively. By the device of place value, a single digit, such as 2, can be made to stand for innumerably different numerical values.

Invention of zero was a result of the need for a device to indicate "no value" for the position where it occurs. Thus, the value of 2 in the numeral 200 is two hundred; without the 0's one would have to guess this from the context in which the numeral was used (which was the case in the earlier Babylonian numerals before a symbol for "no value" was devised). Once a place-value system of writing numbers was invented, invention of a zero symbol was sure to follow — although it could, as in the case of the Babylonian numerals, take several centuries for this to happen (and even longer for the zero to achieve number status).[2] This would also explain the presumably independent invention of a zero in the Mayan culture, which also had a place-value system for writing numbers.[3]

Operations with numerals — addition, multiplication, etc. — were introduced in order to adapt to increasing complexity of social life. They go back, certainly in the case of addition, to the finger-counting stage, and possibly in some instances to the tallying stage. Suffice it to say that they were introduced in response to the same type of stress that forced the invention of the modern computer. The more complex a society becomes, the more efficient must its devices for calculating become.

Study of the history of number evolution also brings out the fact that

[2]See EMC: 61, §2.3 or EMC$_1$: 59, §2.3.
[3]See Menninger, *loc. cit.* Also, for a plausible description of how the Mayan common people operated with numerals, see Sánchez, 1961.

certain identifiable stresses or forces figure prominently in the development of number. (1) *Environmental stress,* chiefly of a cultural character, to which invention of counting devices and operations therewith respond. (2) *Diffusion* from one culture to another; we have already discussed this in the preceding section. (3) *Cultural lag* and *cultural resistance*; again, these have been mentioned in II-4, in regard to the difficulty encountered by the Hindu–Arabic numerals in finding acceptance in Western Europe. (4) *Symboling.* The shifting emphasis, in the evolution of number, on various forms of symboling, with search for more efficient ciphers, and for systems capable of symbolizing numbers in compact form and arbitrary size, brings out the fact that as a culture advances, there evolves along with its linguistic development a special numerical evolution in which symbols play just as important a part as they do in the evolution of its language. (5) *Selection.* In the evolution of number systems, there evidently was a great deal of selection involved. This is apparent not only in the selection of ciphers, but in the bases employed as soon as the systems advanced sufficiently. Although one cannot rule out the possibility of such selection being made by an individual (a priest or other person of authority, for instance), the frequent occurrences of base 10 in differing cultures and its relation to man's ten fingers, suggest that the influence of environmental or culture factors has been more frequently operative.[4]

A special type of selection, which has not been generally emphasized, occurred in connection with the evolution of Greek geometry. The number system of Greece, which was based on letters of the Greek alphabet,[5] played no intrinsic part in the development of Greek geometry. In the latter, so-called "magnitudes" were employed; each line interval could be assigned a magnitude (intuitively its length), relative to a fixed unit magnitude. The employment of magnitudes was apparently forced by the discovery of irrationals, which could not be expressed by the usual number symbols of the Greek culture. The use of magnitudes for ordinary purposes could hardly be expected; for these the Greek numerals were already sufficient and in common use. Magnitudes were part of the geometric (scientific) area of culture, from which the commercial aspects of the

---

[4]For elementary discussion of base and place (or positional) value in the construction of numerals, the reader is referred to EMC: § 4 of "Preliminary Notions;" and for the manner in which special bases were selected in primitive systems, to EMC: Chap. I, § 2.
[5]We refer here to the Ionian numerals; see EMC: Chap. I, § 2.2a.

culture were quite separate. It is interesting to note, however, that a theory of arithmetic can be developed using magnitudes and geometric operations, as was shown by R. Bombelli (cf. Bourbaki, 1960: 168). And although one would hardly expect that such a system would be selected for ordinary purposes of computation, it should be noted that magnitudes are used in modern times on charts and blueprints, as well as in the depiction of statistical data; they are quite generally selected in cases where visual comparison of line segments is more readily comprehended than the corresponding numerical equivalents.

Another well-known and historically important case of selection occurred in the area of symboling for the calculus. The Newtonian "dotage" notation ... $\dot{x}$, $\ddot{x}$, ... for derivatives and $\acute{x}$, $\acute{\acute{x}}$, ... for integrals, competed with the Leibnizian "d" and " $\int$ " symbols. The latter easily won acceptance on the Continent, but the former (presumably due to nationalistic motives stemming from the Leibniz-Newton controversy over the origins of the calculus) continued in use in England until the early 19th century when the "Analytical Society" was formed by Babbage and others to (among other purposes) promote the use of Leibnizian symbols in England.[6]

The importance of a well-selected symbolism is probably nowhere better exhibited than in mathematics (in which we include mathematical logic). This is especially the case where *operations* with the symbols are intended. Symbols like "π" and "e", standing for special numbers, could just as well have been replaced by other symbols.[7] But in the case of the differential and integral symbols of the calculus, the Leibniz symbols lent themselves to efficiency of computation in a way that the Newton symbols could not.[8]

[6]The lag in mathematical research in England during the 18th century is attributed, at least partially, to the refusal to adopt the Leibniz notation, knowledge of which would have revealed to English mathematicians the importance of French and German research during this period.

[7]Although the symbol "π" was first used by the Englishman William Jones, it was probably Euler's use of it that caused it to be accepted by the mathematical community (Jones used it only to denote the 100-place approximation given by John Machin). Euler was also the one to introduce the symbol "e" for the number whose (natural) logarithm is unity.

[8]It might be remarked, too, that important "side effects" resulted from the use of the Leibniz differentials dx, dy. While on the one hand the facility of the "d-operator" in operations of the calculus led one to stubbornly resist giving them up, in spite of the criticisms directed against them, on the other hand the hereditary stress built up in their defense ultimately led to their justification in the non-standard analysis of Abraham Robinson, and a new chapter in analysis. See IV-2 iv-(1).

It is outside the range of the present work to go into the question of just what constitutes a "good" symbol and especially what constitutes a suitable name for a mathematical concept. Apparently a letter from the Greek alphabet (e.g. π) or even less well-known alphabets (e.g. the letter $\aleph$ from the Hebrew alphabet) has an advantage of being more easily associated in one's memory with its designated meaning than the familiar letters of our own alphabet. But selecting operational symbols requires much more skill as well as an intimate knowledge of the related mathematical materials, not to mention the realization that a special symbol will be advantageous; witness the centuries that it took to evolve from so-called "rhetorical" algebra to the modern "symbolic" algebra.

Ideographic symbols were only slowly adopted in mathematics although the need for them was very early recognized in the case of *number* (see Chap. II of EMC). Moreover, adoption of a symbol by an individual mathematician is not any guarantee that it will become a part of the mathematical culture. As might be expected, the status of the individual proposer in the mathematical community usually has an important effect on whether the symbol becomes accepted by the mathematical community and hence part of the mathematical culture. (Recall the remarks about Euler above.) The same is also true of word symbols, or *names,* used to designate either concepts or significant parts of the mathematical culture. And a name once adopted may in the future be rejected for a substitute. For example, the name, *Analysis Situs,* for what we today call *Topology,* was used by B. Riemann in 1851, who attributed it to Leibniz; and this was only four years after Listing's use of the term *Topologie* for his classic *Vorlesungen zur Topologie,* published in 1847. After H. Poincaré's use of the term *Analysis Situs* in 1895 (and thereafter), O. Veblen naturally used the same term in his sequel (Veblen, 1921) to Poincaré's work. The term persisted until the early thirties, when the term *Topology (Topologie)* gradually replaced it, after having been used by such authors as F. Hausdorff, K. Kuratowski and S. Lefschetz.[9]

---

[9]In the first three volumes of the Polish journal *Fundamenta Mathematicae* (1920–3), use of *both* terms *Topologie* and *Analysis Situs* will be found. The advantage of the former term to Indo-European languages is that it lends itself to an adjectival form — *topologische* or *topological.*

3.   Pressure from the environment; environmental stress

As in the case of most subcultures, mathematics has been subjected throughout its history to influences from the environment, and, indeed, owes its very existence to needs of the cultures in which it originated. Counting and measuring systems arise in every culture as it advances. The name "geometry," a Greek derivative, means "earth measure," thereby suggesting social origins. From our recently acquired knowledge of Babylonian history, we know that geometry in Babylon had no status as a special discipline, being more of an accessory to. arithmetic; its function seemed to be partly as a compendium of formulae for calculating lengths and areas, satisfying a social need, and partly as a source of problems in arithmetic. Although the Babylonians apparently knew the Pythagorean formula some thousand years before the time of Pythagoras, it was an isolated feature affording such arithmetic concepts later known as "Pythagorean numbers," which seem to have been studied for their own sake. Likewise, Egyptian "geometry" consisted essentially of formulae for areas and was not a self-contained theory such as one finds in Euclid; it was clearly a tool for a primitive type of surveying.

In Greece, where geometry became a full-fledged discipline, the development of geometry was heavily influenced by philosophy and astronomy. Some of the most influential originators of Greek geometry, such as Eudoxus, were also astronomers. And the consolidation of logic with mathematics, rendering the latter the logically framed discipline found in the *Elements* of Euclid, was apparently in large part due to the influence of philosophers such as Parmenides and Zeno (see Szabo, 1964).

As we approach the modern period and the investigation of physical phenomena following the innovations in science made by Galileo, mathematics was compelled to consider ways of calculating rates of change, speed, acceleration — in general, instantaneous phenomena which the classical mathematics was not equipped to handle. The resulting *calculus* may be said to have introduced the modern era in mathematics, although much work directed toward improvement and perfection of concepts related to the infinitesimal and the infinite, the notion of function and the like, had to be done before the truly modern era of mathematics began.

The study of the theory of heat and sound made by the French scholar

Joseph Fourier during the early 19th century is an outstanding example of the influence of physical problems that emerged during the transition period from the work of Newton and Leibniz on the calculus to the mathematics of the 20th century. Fourier's analysis of the approximations of functions by trigonometric series[10] forced mathematics to generalize the notion of function and as a by-product led to the modern theory of sets.

Although there is a general feeling that mathematics has become, in modern times, more self-sufficient and less dependent upon environmental stresses for its concepts, it is nonetheless true that such stresses continue to exercise a strong influence on the development of mathematics. The impact of World War II, for instance, was instrumental in the invention of more efficient computers and their accompanying theory, as well as in such pursuits as operations analysis, systems analysis, game theory, information theory, not to mention new developments in already established fields.

It should not be forgotten, too, that such fields as probability and statistics, indispensable today for the natural sciences (especially physics), the social sciences and the world of industry, were suggested to mathematicians by their uses in the study of gambling, mortality rates, etc. Here again a typical pattern relating the environment to mathematics ensued:

Environmental stress ⟶ mathematical theory ⟶ environmental applications ⟶ new environmental stresses

The environment suggested the invention of new concepts in mathematics, whose study resulted in mature techniques which were seized upon by environmental interests for the solution of their problems and advancement of their own theories.

4. Motivation for multiple invention; exceptions to the rule

We have already expounded in section 1 of Chapter II the rule concerning multiple invention: When an important new discovery is about

---

[10]Recalling the earlier discoveries of the relations between harmony and the ratios of integers attributed to Pythagoras, M. Kline remarked in his classic *Mathematics in Western Culture* (Kline, 1953), "Whereas Pythagoras was content to pluck the strings of a lyre, Fourier sounded the whole orchestra."

to be made, it will usually be done by several workers independently. Here we consider the reasons for this phenomenon.

"An invention, discovery, or other significant advance is an event in a culture process. It is a new combination or synthesis of elements in the interactive stream of culture. It is the outcome of antecedent and concomitant cultural forces and elements" (White, 1949: 169).

Mathematics is peculiarly adapted to determining what the "cultural forces" are. As a special case, consider an axiomatic system — more explicitly, suppose S is an axiom system for which the resultant theorems are in the process of being discovered and proved. Because of the culturally imbedded methods of proof (laws of ordinary logic, *reductio ad absurdum*, etc.), the theorems of S are essentially fore-ordained. When a given stage in the development of theorems has been reached, the next theorems to be proved are usually discovered (not necessarily in a single order) and become common knowledge in the community of scholars engaged in the development. Given the somewhat uniform capabilities of the members of this community, it is to be expected that proofs of some of these theorems will be found almost simultaneously by several workers in the field.

Similar observations can be made regarding mathematical systems that either have not been axiomatized or which have reached a stage where the theory is so advanced that recourse to the original axioms (which may even have been formulated subsequent to much work in the field) is no longer practicable. The basic theory (Wilder, 1974: 38) of the field has become common knowledge to the workers in the field and operates much as an axiom system in determining future theorems.

We recall our statement in II-1 that the multiple invention of non-Euclidean geometry by Bolyai, Gauss and Lobachewski, as well as the failure of Saccheri to participate in the multiple, can be explained by culturological factors. Consider first the situation of Saccheri. At the time when he worked — the early 18th century — mathematics was still laboring under the philosophy of absolute truth; mathematical theorems either expressed "truths" about the physical or platonic world, or were natural consequences of the traditional theory that formed the basis of "mathematical truth." The geometry of Euclid especially formed a model for this belief. Saccheri, a product of his culture, naturally conformed to this philosophy. At the same time, the geometric vector of the mathematical culture system contained forces stressing for the settlement of the old problems emanating from the Greek dissatisfaction with the parallel postulate (the "flaw" in Euclid's *Elements*). Efforts to prove the postulate throughout the period from the 3rd century B.C. to A.D. 1800 no doubt acted as a further stimulant (stress) on Saccheri to provide, once and for all, a valid proof of the postulate's "truth." His approach, utilizing the classical "Q.E.D." method of the Greeks, was to suppose the postulate false and to obtain therefrom a contradiction. More precisely, he assumed a postulate contradictory to the original parallel postulate, and in deriving its consequences proved an extensive list of theorems concerning what is now recognized as a type of non-Euclidean geometry. But Saccheri could not know this or even conceive of it, in the cultural climate that controlled the thought of his time; an excellent example of the right man being born at the wrong time!

This is not to assert, of course, that Saccheri could not possibly have arrived at the same conclusion that Lobachewski and Bolyai did a century later. Indeed, a stronger willed and more independent thinker who possessed the same mathematical background as Saccheri might possibly have become an example of the "before-his-time" phenomenon (see Chapter VI). However, even if he had realized the true facts, viz. that there could be geometries other than Euclid's, would Saccheri, a Jesuit priest, have had the temerity to publish his conclusions? A century later, the great Gauss, under no threat of ecclesiastical sanctions, did not have the courage to publish his convictions. In a letter to Bessel, written in 1829, he remarked: "It may take very long before I make public my investigations on this issue; in fact, this may not happen in my lifetime for I fear the 'clamor of the Boeotians.'" M. Kline remarked (Kline, 1953: 413–414): He did not have "the moral courage to face the mobs who would have called the creator mad, for the scientists of the early nineteenth century lived in the shadow of Kant whose pronouncement that there could be no geometry other than Euclidean geometry ruled the intellectual world. Gauss's work on non-Euclidean geometry was found among his papers after his death."

One can argue whether the cultural climate of Gauss's day justified such fears. For apparently almost simultaneously with Gauss, Lobachewski and Bolyai reached similar conclusions and published them in 1829–30 and 1832 respectively. That no great hue and cry ensued from the "Boeotians" can be attributed to either of two hypotheses: (1) the cultural climate, permitting greater freedom of scientific thought, had changed sufficiently to allow unorthodox opinions, or (2) neither Bolyai nor Lobachewski had sufficient scientific eminence to warrant their opinions being given serious attention. Probably the truth lies in a combination of (1) and (2). Certainly the cultural climate had changed, as other historical facts would imply, but it was not until after Riemann's *Habilitationschrift* (1854) and Gauss's death (1855) revealed the opinions of these leading mathematicians that the work of Lobachewski and Bolyai received its deserved attention.

There seem to be two exceptions to the multiple invention rule: (1) the unexpected event, and (2) the "before-his-time" phenomenon already discussed to some extent in II-3. In the case of (1), the unexpected event, someone chances upon possible, often paradoxical, examples to which the basic axioms or theory apply. Such cases differ from the usual type of multiple in being outside the orderly sequence of theorems dominated by the basic theory. They can be exemplified by such cases as the Bolzano–Wierstrass discoveries of continuous real functions having no derivatives anywhere; the discovery of paradoxical "biconnected" point sets by Knaster and Kuratowski 1921; or the Banach–Tarski (1924) decomposition theorems. Discoveries of inconsistencies in a theory are of the same type. Cases of these kinds are unlikely to be multiples.

As stated in II-3, the case (2), the "before-his-time" phenomenon, is likely to stem from an unusual combination of background knowledge and experience on the part of the individual creator. In Mendel's case, it was apparently due to his knowledge of bee-keeping and, of course, botany (Iltis, 1966) and in Desargues' case a combination of such architectural details as stone-cutting and a knowledge of classical geometry, especially the work of Appollonius on conics (see Chapter VI).

5. The great consolidations[11]

By a *consolidation* of two or more concepts or theories T_i, we mean the creation of a new theory which incorporates elements of all the T_i into one system which achieves more general implications than are obtainable from any single T_i. The classical example of a new consolidation is analytic geometry (coordinate geometry) in which algebra (and, ultimately,

[11]Consolidation as a cultural force and process will be discussed extensively in Chapter V. Some material will also be found in EMC.

analysis) and geometry are combined, and geometric configurations studied by means of the algebraic or analytic equivalents, and conversely.

Throughout mathematical history, consolidations have taken place. Euclidean geometry, as we know it, is the result of a consolidation of logic and geometry. Until logic was brought in, geometry consisted chiefly of formulae found by experimentation; and, as one might expect, some of these formulae were only approximations (as in the well-known ancient formulae for the area of the circle). With the introduction of logic, proofs by logical inference superceded "intuitive" proofs, and the axiomatic method was a natural result of the process. Some theorems originally not susceptible of proof or extremely difficult to prove by earlier primitive methods became simple to prove by such methods as that of *reductio ad absurdum.*

Another great consolidation that occurred in the Hellenistic culture was attributed to Ptolemy, who combined the remarkable numerical astronomy of the Babylonians with the geometric astronomy of the Greeks to form his planetary system of the *Almagest.* The historian of science, D. de S. Price, has suggested that it was this consolidation that formed the beginnings of modern science (as known in our Western culture) and, moreover, probably the reason why Western science achieved so much in contrast to what other cultures, such as the Chinese, were able to achieve.[12]

Consolidations have become quite numerous in modern times, when a proliferation of specialties has invited them. In fact, this is the way in which mathematics strives to unify its various branches, although the individual motives in any specific act of consolidation are usually the acquisition of tools that will solve problems expeditiously. The angle trisection problem, for instance, was solved by bringing algebra into play, although this was not a case of establishing a new field. The consolidation of algebra and topology that occurred during the first half of the present century was a true example of consolidation between fields, the motivation (on the part of topologists) being to solve problems that had proved too stubborn for the set-theoretic methods previously employed.

It is to be noted, incidentally, that consolidations in mathematics usually leave intact the fields consolidated. In the case of the consolidation just mentioned — algebraic topology — algebra is not only a flourishing field in

[12]Price develops this thesis in his book *Science since Babylon* (Price, 1961).

its own right, but has adopted some of the tools (exact sequences, homology groups) that were created by the consolidated field. Similar observations can be made of differential topology, in which analysis and topology were fruitfully consolidated, and of other modern branches of mathematics.

Usually, fields that have resulted from consolidation are detectable by their names; e.g. algebraic topology, set-theoretic topology, statistical physics. The last mentioned, statistical physics, reminds us that many of the uses of mathematics in its applications are the results of consolidation; the reader can find many such examples if he has a knowledge of any of the modern fields of the natural and social sciences.

6. Leaps in abstraction

In II-6 we discussed the evolution of greater abstraction in mathematics and, to some extent, in general systems (theological, political). Here we will be more concerned with the forces that are operative in such cases.

No doubt the first major[13] leaps in abstraction occurred during the evolution of early Greek mathematics. Unfortunately, the dearth and controversial nature of the history of this period precludes citing any exact dates for their occurrence. However, we can assume rather conclusively that they did occur, if only because of the contrast between the Babylonian–Egyptian type of mathematics and that of the later Greek mathematics. In particular, the Greeks introduced logical proof into mathematics, in contrast to the lack of proof or proof-by-example that characterized Babylonian–Egyptian mathematics.

Recently, A. Szabo (1966) has conjectured that the Greek deductive method was inspired by the Eleatic philosophy, and that the introduction of axioms and postulates as exemplified in Euclid's *Elements* was a consequence of the criticisms of Zeno and the discovery of incommensurables. The latter conjecture has more appeal than others, perhaps, since it provides evidence of cultural stresses which could precipitate the leap to the greater rigor of proof methods.

It is noteworthy that great abstractions have so frequently occurred not

---

[13]The abstraction involved in such achievements as the invention of counting numbers, symbols for them, etc., are too shrouded in pre-history to classify them as leaps in abstraction; most likely they were the results of slow evolution, although this is only conjectural.

as gradual evolutions, but as leaps from previous theories or concepts. This appears sometimes due to mere change of attitude toward existing theories. Consider, for example, the classical problem of the solution of algebraic equations; i.e. the equations of the form

$$a_n x^n + a_{n-1} x^{n-1} + \cdots + a_1 x + a_0 = 0, \tag{1}$$

where n is a natural number $(1, 2, 3, \ldots)$, the a's are integers (positive, negative or zero) and $a_n \neq 0$. The n, we recall, is the *degree* of the equation.

Solutions for the cases $n = 1, 2$ were known to the ancients, even the Babylonians, although usually those solutions which we call negative or complex were rejected. For the cases $n = 3, 4$, the Italian mathematicians of the 16th century (Cardan, Tartaglia, Ferrari) found solutions (see Boyer, 1968: 310–317 for details). Now these solutions were in the form of *algorithms*, i.e. routine methods, or formulae (in a sense, recipes), for finding the solutions. Each algorithm showed how to combine the a's by the operations of addition, subtraction, multiplication, division, and taking roots (radicals) — the so-called *algebraic operations*.

It was only natural that following these discoveries the search for algorithms for the cases $n = 5, 6, \ldots$ should commence. This proved futile; and mathematicians of the 17th and 18th centuries became too involved in the fertile fields of analytic geometry and calculus to spend much time on the problem. Nevertheless, the gain in mathematical power and maturity during these centuries presumably contributed to the change in attitude that led to the solution. In particular, the French mathematician Lagrange, whose breadth of interests in addition to his work in analysis and its applications made him famous, became interested in the solution of algebraic equations, publishing in 1767 a paper on solutions by means of continued fractions and in 1770 a paper in which he considered solutions in terms of permutations on the roots. But his greatest step forward was to ask, "Why do the methods, which are successful in solving equations of degree at most 4, all fail when the degree is greater than 4?" and conjectured that algebraic equations of degree greater than 4 are not solvable by such methods.

Even to ask such a question represented a great leap in abstraction. The question was clearly forced by the internal stress which was generated by the frustrations created by the unsuccessful attempts at solutions. And it was the Italian physician Ruffini who, in 1813, tried to prove Lagrange's conjecture that the general equation of degree greater than 4 has no general solution using only algebraic operations.

The sequel, viz. work of Abel and Galois, which culminated in the theory of substitution groups (which in turn was one of the stimuli for work in the general theory of groups), is well known. Moreover, from this time on, what are usually termed "structural" aspects of mathematical phenomena, as exemplified in the intense research into the structure of abstract groups, became a prominent feature of the evolution of mathematics and a decisively important factor in the passage to modern mathematics.

Although we are presenting here only certain selected episodes from the history of mathematics, and especially not an exhaustive account of the history exemplifying leaps in abstraction, there are two more modern episodes which we feel would serve to further clarify the process. The first of these, *theory of sets,* was introduced during the latter part of the 19th century by Cantor and Dedekind, as a tool for the solution of function-theoretic problems. Its extension, by Cantor, into a realm of transfinite numbers was a tremendous leap which stirred up controversy among mathematicians despite the proved usefulness of these numbers. Indeed, their fruitfulness and intrinsic appeal led to their ultimate acceptance by the majority of the mathematical public in spite of the paradox and contradiction resulting from their unrestricted use. This is an outstanding case of the effect of internal stresses in mathematics forcing the adoption of ideas which would otherwise have never been proposed, much less accepted. By the time contradiction appeared, the theory of sets had built up sufficient stresses within itself to stimulate the investigation of means of so eliminating the contradictions as to permit the continued use of the theory in a consistent manner.

Finally, we wish to mention a leap in abstraction that resulted from the consolidation of the theory of sets with mathematical logic. It should be explained that the logic traditionally associated with the name of Aristotle and later forming a basic study in Middle Age philosophy, especially of the Scholastic School, was converted to an algebraic form, using the tools of the by then available abstract algebra, by the self-taught English mathematician George Boole (1847, 1854).[14] Later classics in the field of

[14]Note that this was in itself a form of consolidation, viz. of logic and algebra.

the resulting mathematical logic include Frege's *Die Grundlage der Arithmetik* (1884) and the Russell–Whitehead *Principia Mathematica* (3 vols., 1910–1913).[15] It was principally the attempt of D. Hilbert and his students during the first quarter of the present century to subject the entire field of mathematics to logical scrutiny, mainly to prove its consistency when based on suitable formal logical rules, that created a new level of abstraction, a "metamathematics." Since the basic motivation for this was the inconsistencies found in the theory of sets, this metamathetics involved a consolidation of the theory of sets with mathematical logic.

Hilbert's attempt was shortly rendered futile by the work of the young Austrian mathematician K. Gödel, who made the epoch-making discovery in 1932 that no formalization of mathematics in terms of mathematical logic can ever be complete so long as it is consistent.[16] The sequel was a host of discoveries, which cleared up a number of unsettled questions in set theory and in turn created many new unsolved problems.

One might predict from the developments just described that mathematics has reached the height of abstraction. This is not to imply, however, that individual fields of mathematics, as for instance computer theory, cannot within themselves become more abstract. The achievements of mathematics as a whole (including mathematical logic, of course),[17] bear witness to the aphorism that increased abstraction generates increased power. Only the future can reveal whether mathematics as a field can achieve higher levels of abstraction than those now attained, or whether it will be only specialized subfields of mathematics that will be able to make further leaps in abstractions.

In conclusion, we might observe that in view of the historical fact that as any scholarly theory evolves, it becomes more abstract, abstraction is one of the most effective forces prevalent in the evolutionary process.

---

[15]Strictly speaking, each of these works was designed to found mathematics on basic principles of logic. But they served to help construct the type of symbolism later to become standard in mathematical logic.

[16]Actually, Gödel's result referred to any formalized mathematical system broad enough to contain a formal counterpart of elementary number theory.

[17]In earlier times, logic was considered as a subfield of philosophy; today, however, mathematical logic is usually considered as a field of mathematics.

7. Great generalizations

Closely related to leaps in abstraction are the great generalizations made in mathematics. In fact, generalization early became one of the chief tools for accomplishing abstraction. In EMC: 40 ff. (EMC_1: 39 ff.), for instance, we noted the examples of primitive numbers that were specialized to certain categories of objects. For example, in the Tsimshian culture of British Colombia, number words for men differed from number words for canoes, and a modern form of the phenomenon persists in the Japanese number words in the range from 1 to 10. To pass from such number words to "pure" number words of a non-categorical form requires an evolutionary form of generalization. An analogous form of generalization occurred in early geometry, when special cases of areas found, for instance, in the Babylonian–Egyptian mathematics were replaced by the general forms characteristic of the *Elements*. The historical details of how these generalizations evolved are unknown.

We know more about the generalization that carried special numerical operations over into what we ordinarily term "school algebra." This was not a leap, but occurred slowly over many centuries, passing through so-called "stages;" the "rhetorical" stage, the "syncopated" stage, and the "symbolic." The rhetorical stage is usually illustrated by examples found on Babylonian clay tablets dating from as far back as 1900 B.C. and employed explicit verbal statements involving the sexagesimal numbers used in that era. The syncopated algebra employed both abbreviations and words; e.g. K^γ for the cube of the unknown, derived from the word KYBO (Diophantus, early Christian era), and the word *aequalis* for "equals" (Cardano (1501–1576)). There was a progressive evolution of symbolism until the 17th century when algebra, essentially as we use it in the high schools and elementary algebra courses of the colleges, appeared.

We can observe here how generalization serves to achieve abstraction. Certainly, as any high school student beginning the study of algebra can bear witness, "word problems", i.e. problems stated entirely in words, are more comprehensible than the equivalent algebraic formulation which is introduced to achieve the solution. The latter is more abstract, and the average pupil encounters his greatest difficulty in converting the "word problem" into the equivalent algebraic form. However, he usually finds the latter simple enough once he has acquired the suitable algorithm for its solution — although he may then encounter difficulty in interpreting (in words) the solution offered by the algorithm. Incidentally, this example affords additional testimony of the power of abstraction (in the form of the algebraic equation).

A unique type of generalization is achieved by the process of abstraction. As summed up by the pioneer American mathematician E. H. Moore, the process is based on the principle: "The existence of analogies between central features of various theories implies the existence of a general abstract theory which underlies the particular theories and unifies them with respect to those central features" (E. H. Moore, 1910) (see also EMC: 175; EMC_1: 169). The classic example is that of the evolution of group theory. Here again no leap occurred. Although Galois originated the term "group" apparently, he did not originate the *concept* of group. The latter arose gradually in various forms, through recognition of similarity in operations occurring in arithmetic, algebra and geometry. A recognized definition of group did not evolve until the 19th century; and with the rise of abstract algebra, the notions of *fields, rings, commutative algebras,* etc., evolved.

Another, and associated generalization, must be mentioned, viz. the classification of geometries by Felix Klein in his *Erlanger Programm* (1872), by means of their underlying transformation groups. Although no longer furnishing a complete definition of geometry as it exists today, the *Programm* was a fundamental contribution at its time, and illustrates the manner in which E. H. Moore's more recently stated principle can contribute to generalization.

One of the most celebrated generalizations of recent times is the well-known Atiyah–Singer index theorem. It generalizes the classical Riemann–Roch theorem of analysis and consolidates materials from both modern topology and analysis to show that the "analytic index" can be computed solely from topological data. The details are too technical to include here.

As a tool for research, generalization is one of the most important tools of the research mathematician.

CHAPTER IV

Potential of a Theory or Field; Hereditary Stress

A great department of thought must have its own inner life, however transcendent may be the importance of its relations to the outside.

E. W. Hobson

Anthropologists have devoted much work to the study of *cultural change* or *evolution of culture*. A principal value of such studies lies in their contribution to the art of prediction; having observed how cultures have changed in the past, one may with greater confidence conclude what to expect in the future.

Change in the mathematical culture consists chiefly in the introduction of new concepts. Where do such new concepts come from? To answer, "from the minds of individual mathematicians," would be both true and false; true in the sense that cultures work through the minds of their carriers, but false in that it is the individual's accumulation of previously invented concepts derived from his culture that motivates his own thinking and leads to new concepts. It is frequently asserted that it is the *intuition* of the individual mathematician that lies at the heart of his conceptual thinking; but here again one ignores the fact that the individual intuition is itself a result of the culture's effect upon the individual's mind.[1]

Just as a vector from the United States political conglomerate exerts its influence through the individual lobbyist (White, 1975) so, too, do mathematical vectors achieve conceptual realization through the thinking (intuitive, overt expression of previous mathematical concepts) of the individual mathematician. The latter's conclusions are a direct result of the synthesis activated in him by his particular mathematical culture. It is to the mathematical culture that we turn, then.[2]

[1]Regarding intuition see Wilder, 1967; intuition will also be discussed in Chapter 7 in connection with law 9.

[2]One interested in the manner in which the individual mathematical mind works is referred to Hadamard, 1949 as well as to Poincaré, 1946

1. Hereditary stress

The most important force exerted from the mathematical culture upon the individual mathematical mind we have labelled "hereditary stress," for the reason that we inherit it from the previous, already existing, state of the mathematical culture.[3] The stress is not only responsible for our impulses toward new concepts, but is frequently compulsive in nature — in the sense that unless we invent new concepts or accept concepts hitherto rejected, we shall be unable to accomplish our purposes. One may recall the concepts born out of the necessity of accepting complex numbers if we are to have a satisfactory theory of equations; or the acceptance of computers in mathematical thinking, just as the Greeks accepted ruler and compass as part of their geometric thinking.

Hereditary stress is not a simple concept, since one finds, upon analysis, a number of components involved in its expression as a force for change. In other words, hereditary stress is a *collective* term, embodying all those characteristics of a field or theory that influence it to growth. Of course one can ask, why should it grow? History gives many examples of prominent mathematicians who arrived at a state where they concluded that mathematics could not grow much further: Lagrange, for example, came to believe that mathematics as a whole was near exhaustion. Charles Babbage, one of the most prolific of 19th-century mathematicians, asserted in 1813 that, "The golden age of mathematical literature is undoubtedly past." Yet despite such pessimism, greater days for mathematics were ahead.

Attainment of what might be called a stable equilibrium of cultures is not unknown; the aboriginal Australian culture is one of the favorite examples of anthropologists. But there is a difference between such cultures and a scientific culture, and particularly the Western mathematical culture upon which we are concentrating. The latter, unlike the early Chinese mathematics or, for that matter, the Greek and Hellenistic mathematical culture, has attained a state where continual growth seems assured, barring catastrophe from without. Presumably an antagonistic host culture can be fatal to scientific growth; in both the Chinese and Hellenistic cases just cited, this was likely the case. Such cases are illustrative of extreme environmental stress. From the time of Pythagoras (*ca.* 540 B.C.) to that of

[3]Our article Wilder, 1974 was devoted to a study of this stress; the present chapter represents a revision, updated, of this article.

Diophantus (A.D. 250) there may well have been built up a sizeable hereditary stress in mathematics, but the closing of the school at Athens (A.D. 529) was not far off. Plants do not grow in poor soil and climate. And as we shall see, even an internal force such as hereditary stress is not immune from environmental influences.

A situation has been reached today where most of the subfields of mathematics whose birth and growth were instigated by environmental factors (e.g. statistics, probability, operations research, computer technology) have seceded from the parent that fostered them and taken up abodes of their own in the academic community. A number of these, such as operations research, resulted from the demands made on mathematics in World War II. Their retention in the departments of mathematics would no doubt have led to financial and administrative burdens that would ultimately have made a virtual pauper out of the mathematical core (the so-called "pure mathematics").

It will be our purpose to restrict our study of hereditary stress to this core, consisting of analysis, geometry, topology, algebra — in short, those studies that constitute the basic scientific mathematics. It is important to recall that these subfields are not sharply cut off from one another; rather they have borrowed from one another and merged many of their theories so that they form today a consolidated whole — the *core* of modern mathematics. From the standpoint of mathematics as a cultural system (I-4), hereditary stress can be regarded as the attractive force between vectors which not only causes their growth but assists in their merging or, possibly, absorption of one by another. In the following analysis, we try to select those properties of a theory which contribute to hereditary stress.

2. Components of hereditary stress

What, specifically, are those characteristics of the mathematical culture that contribute to growth? We shall not pretend that we have found them all, but the following seem to be the major ones.

(i) Capacity

This is, roughly speaking, the potential of the theory to produce a meaningful and fruitful mathematical yield. A classical example would be the capacity of a complete set of axioms for Euclidean plane geometry. Such a set would have, initially, great capacity, in that it would presumably lead to an unending series of theorems. It must be noted, however, that in speaking of the capacity of an existing theory — one viewed from the *present* — the capacity usually differs from that of former states. Today, the capacity of plane Euclidean geometry is small; the field is well worked out, at least from the standpoint of what it can contribute to mathematics as a whole. This does not mean that there are no theorems yet to be discovered; and, indeed, interesting and important ones. In theory, infinitely more theorems could be stated than are presently known, although it may be more and more difficult to elicit them.[4]

Capacity, then, is dependent on the time at which we view a field. A modern analogue of what we have said about the capacity of Euclidean plane geometry is that of the topology of the Euclidean plane. This was well worked out, particularly by the R. L. Moore school in the United States and the Polish school (Sierpinski, Kuratowski, Mazurkiewicz and others) during the first part of the present century. Its present capacity seems to be much like that of Euclidean plane geometry. On the other hand, the capacity of topology, including all its subfields such as general topology, algebraic topology, differential topology, etc., is still apparently large. Consolidations with other fields in this case have resulted in enormous increase of capacity.

The trained mathematician can usually sense the capacity of a given field. If the capacity is large, he may either be attracted to do research in it, or direct others, such as younger colleagues or students, to consider helping to develop the field. In this way, capacity exerts a direct effect on the growth of a field.

However, it can happen that even the most experienced and informed mathematicians, at a given time, will fail to recognize the capacity of a theory. Although it will be discussed more fully later (Chapter VI), a good example is the start in what we today know as "projective geometry" made by Desargues in the 17th-century. Presumably Desargues's contemporaries failed to recognize its capacity, as indeed did Desargues himself, and failure to develop it was probably due (as will be explained later) at least partially

---

[4]For an up-to-date sample of geometry, not confined to the plane, including a description of some of the new methods that have been introduced since the time of Euclid, see Coxeter and Greitzer, 1967. Also see Klee, 1979.

to the state of the geometric vector at the time; it consisted mainly of Euclidean geometry and the newly consolidated "coordinate geometry."

Recognition of the capacity of set theory by both Dedekind and Cantor during the latter part of the 19th-century resulted in a new vector (theory of sets) stemming from the analytic vector of mathematics. This is an outstanding example of the manner in which a new field may be born. It was Cantor, of course, who persisted in the belief that extended study of the transfinite and its arithmetic could stand on its own; others were slower to recognize its capacity, and had doubts concerning its "existence."

Following the work of Hilbert and his students regarding formal systems, and the discoveries made by Gödel and Paul Cohen concerning the continuum problem, the capacity of set theory was further revealed. This was a kind of renewed or rejuvenated capacity, stimulated by consolidations with symbolic (mathematical) logic. The capacity of the resulting vector is currently growing not only through its own internal development, but in its implications for other disciplines, both for philosophy and for the study of social phenomena.

Another example was the recognition of the capacity of the topology of the plane and of the structure of continua by R. L. Moore and the Polish school of W. Sierpinski cited above. An adequate study of this development and its influence upon later mathematical history is unfortunately still lacking.

(ii) Significance

This is a quality narrowly related to the capacity of a theory, and like capacity may vary greatly with time. Its inclusion as a component of hereditary stress results from the fact that it is such a strong factor in attracting new investigations, and hence in increasing the potential of the field vector.

For an example, we may again cite the case of set theory. For some time its significance for mathematics was not generally realized; some even denied its significance *in toto.* Presumably those mathematicians interested in the development of theories of integration (e.g. Borel, Lebesgue) were somewhat sensitive to the inevitability of investigations into the theory of sets, but true realization of the significance is a mark of a 20th-century mathematics.

Of course, significance is not necessarily to be weighed by the importance of a field for mathematics alone. Just how significant for mathematics is computer theory, for example? At present (1979), mathematicians who gave themselves over to developing the computation vector of mathematics may likely find themselves no longer in the mathematics departments of the universities but either in electrical engineering schools and/or as constituting departments in their own right. On the other hand, the growing usage of computers in the teaching of calculus and higher forms of analysis, as well as the use of the computer in solving the famous Four Color Problem[5] argue for the continued significance of computer theory for mathematics. Those who object to the introduction of a material entity (the computer) into mathematics may be reminded that even the "pure" Euclidean geometry admitted the concept of ruler and compass.

Significance for fields external to mathematics is not to be ignored, however. Growth of a mathematical vector may be due not only to mathematical needs, but may even be motivated by other influences. Graph theory, for example, was originally used by Kirchhoff for the investigation of electric circuitry and by Cayley for the study of chemical bonding long before being recognized as a chapter (one-dimensional complexes) of topology. Today the graph-theoretic vector has grown so significant that it is not generally thought of as a subfield of topology, but as a field in its own right. Some of this, to be sure, is due to its significance to other fields than mathematics (e.g. physics, social sciences); but as a self-sufficient field it is rightly included within the framework of pure mathematics.

In contrast to the above cases, significance may be a *decreasing* feature of a field. We have already mentioned above the decreasing capacity of Euclidean plane geometry; concurrently the significance of this field seems to have decreased. In this, however, it is possible that a tragic mistake has occurred. It is generally granted that some mathematicians are *visual* thinkers and others not; some visualize their work in geometrical patterns. Emil Artin, certainly one of the world's greatest algebraists of the 20th-century, was a visual thinker; some of his most important work, in fact, was done in topology (as, for instance, his classification of braids). A geometer who listened to him lecture soon recognized a geometric pattern in the way Artin organized his presentation of algebraic concepts. This lead, I think,

---

[5]See Appel and Haken, 1977.

to a precaution concerning mathematical education, viz. that geometry, particularly of the Euclidean type, has a profound significance for the developing mind, not necessarily for its exemplification of the deductive method, but for the visual explorations involved.

The renowned Hilbert remarked in his famous paper on mathematical problems (Hilbert, 1901–2),

> ... geometrical figures are signs or mnemonic symbols of space intuition and are used by all mathematicians. Who does not always use along with the double inequality $a < b < c$ the picture of three points following one another on a straight line as the geometrical picture of the idea "between"? Who does not make use of drawings of segments and rectangles enclosed in one another, when it is required to prove with perfect rigor a difficult theorem on the continuity of functions or the existence of points of condensation? Who could dispense with the figure of the triangle, the circle with its center, or with the cross of three perpendicular axes? Or who would give up the representation of the vector field, or the picture of a family of curves or surfaces with its envelope which plays so important a part in differential geometry, in the theory of differential equations, in the foundation of the calculus of variations and in other purely mathematical systems?

The significance of a field may be due to any of a variety of factors; its significance for other parts of mathematics, to fields external to mathematics such as physics, or to mathematical education. During the pre-World War II era, at least in the United States, the significance of mathematics as a whole was dangerously played down by professional educators, presumably on the grounds of theories regarding "transfer" of abilities. However, the significance of mathematics to the military effort proved a deterrent to the waning participation of mathematics in the school curricula. The boost given mathematics at that time seems to have lasted to the present time (1979) although some observers report signs of decreasing interest in mathematics on the university level.

(iii) Challenge

By this term we refer to the challenge created by the emergence of problems which seem to defy solution or to require unusual ingenuity or new methods and principles for their solution. Such a challenge affects not only workers in the field concerned, but may — and usually does — attract workers from without the field. The famous problems presented by Hilbert at the 1900 International Congress of Mathematicians in Paris (*loc. cit.*) are an excellent example of outstanding problems in various fields which have attracted workers ever since.[6]

Consider, for example, one of the oldest fields in mathematics, viz. number theory. Granted that simplicity of the original basis — the natural numbers — and the introduction of new methods from analysis and algebra have served to preserve its popularity, even among amateurs; but a major role has been played by the challenging problems to which number theory has given rise.[7] It is true that number-theoretic results have proved useful to other fields of mathematics, but it is its fascinating problems that have kept the field alive. As Hilbert stated (*loc. cit.,* p. 438), "As long as a branch of science offers an abundance of problems, so long is it alive; a lack of problems forecasts extinction or the cessation of independent development. Just as every human undertaking pursues certain objects, so also mathematics requires its problems." The challenge offered by problems may justifiably be considered as one of the most important components of hereditary stress.

(iv) Conceptual stress

In II-7, we have already discussed this force to some extent, observing its effects in the domain of numbers, geometry and theory of sets. Here we extend our previous remarks.

One of the most important components of hereditary stress, created by the need in the field concerned for new conceptual materials, conceptual stress operates in several ways:

(1) Symbolic stress. This occurs in two forms. The stress resulting from the need of a good symbol or symbols; we have already discussed this in III-2. Then there is the reverse of this — given a "symbol," find a suitable meaning for it. From the standpoint of the symbolic process as usually conceived, this is nonsense. By definition a symbol stands for something, so

---

[6]For an up-to-date discussion of these and the solutions which have been obtained, see Browder, 1976.

[7]Number theory is another field of mathematics which has benfited from the invention of the computer. For example, it has made possible the discovery of new perfect numbers (discussed in Book IX of Euclid's *Elements*); see, for instance, Tuckerman, 1971, announcing the determination of the 24th even perfect number.

why raise a question regarding what a symbol stands for? Actually, the symbols we refer to here do stand for something, except that their "reality" is open to question. By the word "suitable" above, then, we mean an accommodation to this "reality." Some examples will perhaps make this clear.

Consider the classical case of $\sqrt{-1}$, already discussed in II-7. As a symbol, it stands for "the square root of -1." Unfortunately at the time the symbol was first encountered, the only numbers known were of the "real" type — what we today term "real numbers."[8] It was only after further cases where such "numbers" forced themselves to the attention of mathematicians — particularly when, as already stated above, it was discovered that the fundamental theorem of algebra could not be stated without the admission of "imaginary numbers" — that genuine effort was made to provide them with a proper conceptual background — in the role of "complex numbers" — along with justification for modes of operating with them. Symbols, as a means of thinking, interact and generate new symbols!

Another elementary example is provided by the Leibnizian d-symbols, dx, dy, etc., mentioned earlier in III-2, in connection with our discussion of *selection*. These had to wait nearly three centuries for a rigorous conceptual meaning. In the meantime, because of their marvelous adaptability to operational techniques, they survived both philosophical and mathematical assaults. No better illustration can be given of the persistence of a good mathematical symbolism. They "worked," so why worry about a *meaning* for them? Of course, teachers of mathematics like to make their ideas seem "logical" — *rigorous* is the usual word — and in the attempt, many and wondrous have been the meanings associated with these "little zeroes" dx and dy. For the associated symbol dy/dx, denoting the derivative, both Cauchy and Bolzano (*ca.* 1823) gave satisfactory definitions in terms of the limit idea.[9] But texts in both calculus and advanced calculus, striving to preserve an admittedly valuable instrument,

---

[8]Not *all* types of real numbers were conceived at that time (*ca.* 1545), to be sure. Cardan, whose *Ars Magna* contained the solution of the general cubic (in which the so-called "imaginaries" first forced themselves to mathematical attention), even called negative numbers *numeri ficti*.

[9]This was another multiple, not mentioned in our previous discussion of such cases. (Boyer, 1968: 564–565, terms it "only a coincidence.")

used various devices to give conceptual justification for the *differentials* exemplified by d*x* and d*y*. The most successful of these was probably the treatment of "d" as an *operator,* a notion due especially to G. Boole, whose work on the calculus of finite differences is known to all professional actuaries.

A satisfactory conceptual treatment of the differential was given by A. Robinson, who availed himself of modern set theory — another case of a theory having to wait for the development of a new field — theory of sets — before it could be given a satisfactory meaning.[10]

From a general point of view, it is well known that the entire history of mathematics is marked by an almost continual stress for greater and more suitable symbolism. Increasing abstraction and complexity have forced this. Of course, the path has not always been smooth; recall the differences between the "synthetic" and the "analytic" school of geometry in the later 18th and early 19th centuries.[11] Although the latter was a quarrel over methods, it had a symbolic basis in that the one (synthetic) preferred symbols of a linguistic form whereas the other insisted on the superiority of the literal and ideographic (see III-2) symbols and operations therewith that had by that time become characteristic of algebra and analysis.[12]

(2) Problems whose solution requires new concepts. Another type of conceptual stress occurs in connection with problems whose solution can only be effected by the creation of suitable new concepts. The solution of the problem of incommensurables and radicals in Greek geometry is attributed to Eudoxus's creation of the theory of proportionality as described in Book V of Euclid's *Elements.* The analogous treatment of the real continuum by the late 19th-century analysts was again created to meet the problems associated with the arithmetization of analysis. Also attributed to Eudoxus is the creation of the "method of exhaustion" which

---

[10]Robinson's work may be found in his book *Non-Standard Analysis* (Robinson, 1966). For an example of how the non-standard analysis may be used to found the calculus, see the textbook Henle and Kleinberg, 1979.

[11]See, for instance, Boyer, 1968: 578–579.

[12]Recall the remarks made above in our discussion of *Significance* concerning visual and non-visual thinkers. Presumably this may have been partially responsible for the debate, although both Monge and Poncelet were quite at home in both analytic and synthetic geometry.

solved the problem of defining lengths of curves and areas bounded by curves, etc.

The problem of the solution of the general algebraic equation was completely resolved by the theory of groups and Galois theory. And the problems raised in the foundations of analysis, especially of integration, were solved only by the creation of the theory of sets by Dedekind and Cantor, and the invention of measure theory by such investigators as Borel and Lebsegue.

Often the creation of new concepts leads to considerable opposition, as was exemplified in the antagonism to Cantor's theory of the infinite. The concepts of Intuitionism, stemming from Kronecker, but as a system created chiefly by L. E. J. Brouwer, as a solution for the contradictions in the theory of sets, met such strong opposition that the tenets are accepted only by a small minority of the mathematical public today. That Intuitionism was a consistent theory was generally admitted, but it implied such radical surgery on existing mathematics that it met general rejection.

However, not all such proposals meet such opposition. Presumably Eudoxus's theories were welcomed, and Galois theory is a standard theory in algebra today. That Cantor's theory of the infinite met with such opposition seems to have been due not merely to the contradictions it engendered, but to a general cultural block in both mathematics and philosophy exemplified by Gauss's dictum, "Infinity is merely a *façon de parler*." Its usefulness in modern mathematics, in its modified forms (Zermelo–Fraenkel, von Neumann–Bernays), seems likely to overcome the opposition which still exists.

(3) Stress for creating order among alternative theories. It can happen in mathematics that a theory gives rise to alternative theories each of which has its advantages, but whose accumulation ultimately leads to frustration and confusion. The stress toward introducing order in such cases usually leads to the creation of new conceptual frameworks in which the individual theories take a natural place. Klein's *Erlanger Programm* was a classical example; analysis of existing geometries led him to realization of the importance of transformation groups and geometrical invariants as their distinguishing features. The resulting clarification of geometries served both to introduce order into the field of geometry, and as a stimulation to

research, at least until the space–time concepts of relativity led to broadening the base of geometry in general.

The group concept was itself a unifying stimulus in the mathematics of the late 19th century and early 20th century. The evolution of the notion was quite gradual, not emerging into a precisely formulated definition until nearly a century after its early occurrences in such work as that of Galois. Realization of its universal influence in mathematics led eventually to its formalization in axiomatic form and to its establishment as a recognized field of mathematics.

Of a more specialized character, but rapidly increasing importance in mathematics despite its recent birth, is the category theory of Eilenberg and MacLane. We mention it here since its original motivation was to classify and identify the various homology theories that had been devised in the algebraic topology of the first half of the 20th century. Incidentally, the contrast in the time that it took for this theory to reach such widespread recognition, and the time consumed in the earlier formalization of the group concept, bears witness both to the increased homogeneity of the current mathematical culture and especially to the universal recognition therein of the efficiency and utility of consolidating broad conceptual devices in precise axiomatic form.

(4) New attitudes toward mathematical existence. Finally, and on a grander scale, there is the conceptual stress promoting new attitudes toward *mathematical existence* and so-called "mathematical reality," matters bordering on the general philosophy of mathematics and the foundations of mathematics. The question of what constitutes mathematics has never been satisfactorily settled.[13] Of course, the times when there were different mathematical cultures (e.g. the ancient Chinese, the Mayan, the Arabic, etc.), if it is proper to call them such, has passed away. One may possibly still be able to point out differences in national mathematical interests, but these are minor characteristics of the modern mathematical culture. The mathematical communities of Western Europe, the U.S.S.R. and the United States may be said to form the dominating

---

[13]Anthropologists unable to agree on a definition of culture can point to mathematicians and their predicament in this regard.

cultural core of present-day mathematics; its influence on the other cultures such as those of Asia, South America and Africa exercises a beneficial diffusion contributing to a much desired world-wide unity, without sacrifice of national interests.

But even during the European Renaissance period, what was considered mathematics had a much wider range than would be admitted today, especially regarding applications. At the present time, due especially to the expansion of new vectors (cf. I-4), what is considered the core of mathematics is much more specifically defined. This does not mean that so-called "good" mathematics is confined to the departments of mathematics, although it is generally confined to the universities and technical laboratories. The "free-lance" period, during which such men as Fermat, Pascal, Descartes and others pursued mathematics as a hobby or side line, is essentially gone. Although there are many "amateurs" today, their work is not comparable to that of the great amateurs — presumably because the mathematical culture has reached such a state of abstraction; only the professional (the "technician") is, as a rule, capable of significant creative work.

It is not surprising, in view of the above remarks, that what is considered *mathematics* has varied both from culture to culture and from one individual mathematician to another. But these conditions do not imply that all was, or is, chaos in mathematics. Despite differences of opinion, certain mathematical doctrines are usually agreed upon. The classic example is probably the geometry of the Greeks as codified in Euclid's *Elements*. For roughly two and a half millenia, the *Elements* was considered as embodying "mathematical truth," its axioms and postulates so "obviously true," that to question a single one risked condemnation. True, the parallel postulate was suspected as being a logical consequence of the other postulates, but no one doubted its "truth."

With the introduction of non-commutative algebras and non-euclidean geometries in the 19th century, the stress toward new conceptions of mathematics led to the recognition that, as Pioncaré put it, mathematics "is a free invention of the human mind;" which, properly interpreted, is a true description of the situation at the end of the 19th century. It expressed a doctrine whose essence was that mathematics is not bound by physical reality or by the concept of absolute truth. This was a concession necessary to the acceptance of conflicting algebras and geometries; from then on, "truth" in mathematics could not be absolute, but only relative to clearly set forth basic assumptions.

> This new attitude contributed enormously to the creation of new mathematics, although it did not authorize unrestricted freedom. Freedom to introduce new theories remained tied to the existing mathematical culture; not until the latter had been augmented by new attitudes toward what is permissible in mathematics could the requisite "freedom" be exercised. No longer need mathematics justify itself by its so-called "reality." The new freedom allowed one to introduce new conceptual materials wherever they would prove of significance and contributory to the currently evolving conceptual body of mathematics. This was a freedom having a temporal character, in that it was modified by the current needs and interests of mathematics. What is not significant in one period may turn out to be so in a later period. This, incidentally, contributes to an explanation of the historical "singularities" that occur in the sciences — the inventions that occur "before their time." History provides many examples of credit for a "discovery" given to an individual whose good luck it was to introduce it at the time when it was most needed — precisely at the time when it became most significant — although historical research later reveals that the discovery was made earlier by someone else.
>
> Acceptability of new conceptual materials plays a related role of course, in that such materials may not be acceptable at one time yet find no difficulty achieving status later as cultural attitudes change. It can be surmised that Saccheri might have achieved credit for the invention of non-euclidean geometry if the mathematical culture of his time had advanced to the level which nineteenth-century mathematics reached; the invention of new algebras, for instance, was already acceptable. As it was, he "settled" for a spurious *"Euclides Vindicatus"* which conformed perfectly with the views constituting the mathematical culture of his time. And even though the nineteenth-century viewpoint had been modified to the point where the non-Euclidean geometries might achieve recognition, this was not accomplished without a struggle.[14]

Inevitably, along with the new "freedom" there remained sectors of mathematical opinion that cautioned restaint. Outstanding were the beliefs (mentioned above) of Kronecker, whose insistence that mathematics should be based on the natural numbers and built thereupon only by finite constructive steps did not find a welcome reception until the early 20th century, when they were embraced by L. E. J. Brouwer as constituting a philosophy free from contradiction such as the new theory of sets had engendered.

More recent alternatives in the form of "constructive mathematics" can be exemplified by the work of E. Bishop.[15] Such theories seem to have found a ready acceptance in the philosophy of mathematics, but if

[14]See the two middle paragraphs of p. 44 of Wilder, 1974.
[15]See Bishop, 1967.

mathematics were represented by a vector system based on philosophical opinion, the "constructive" vector would be relatively small. The hold of the platonic philosophy on mathematics would still be represented by a vector of relatively large magnitude. Moreover, the more recent justification and expansion (even into current calculus textbooks, as cited in (1)) of the Leibnizian techniques seems to be a strong countervailing vector; it is presumably acceptable to the platonic point of view.

(v) Status

Of a cultural nature, and contributing heavily to any vector representation of a field within mathematics, is the esteem in which the field is held among mathematicians, and which we term "status."

Although presumably a stituation which is not well known among the non-scientific public, the fact is that mathematicians frequently have taken strong opinions regarding the status of the various fields of mathematics. One may expect that the individual mathematician has high regard for his main field of interest, although history records changes of attitude under the influence of philosophical criticisms.[16] Kronecker, already cited above for his radical ideas regarding the nature of mathematics, while prominent for his fundamental contributions to algebra, held many modern mathematical ideas (especially those of Cantor) in great contempt. Also, cleavages between so-called "pure" and "applied" mathematicians have been frequent, and we have already cited the antagonism between proponents of "synthetic" and "analytic" methods during the early 19th century.

Opinions other than philosophical affect the status of a field, however. Of especial importance in the contributory factor — how much the field may contribute to other fields of mathematics as well as its possible contributions to the neighboring sciences, both physical and social. Algebra, both ancient and modern, has gained much of its high repute from its important uses both within and outside the mathematical core, and a similar statement holds for mathematical analysis. Topology, in the so-called "general" form consisting of the theory of topological spaces, has

---

[16]Such was the case especially among set-theorists who became convinced by anti-set theoretic criticisms; cf. especially Lusin, 1930: 27.

gained a strong position as a foundation for analysis, and to some extent in modern algebra. The status of geometry is probably not quite as high as it was during the Renaissance, but this pertains only to its function as a field for research. As a background and foundation for both the mathematician and the engineer, as well as the physical scientist, geometry retains its significance and status.

The concept of status is important chiefly because, with increased status, the attraction of a field for creative mathematicians also increases. Moreover, graduate students who notice the increasing status of a field are more likely to be attracted to it, as are also workers in bordering fields. An obvious indicator of the status of a field is the number of research mathematicians involved in its development, suggesting that a "worker ratio" consisting of the number working in the field relative to the total number of creative mathematicians could serve as a measure of the field's status. However, an easier ratio to compute would be the ratio of papers in the field to the total number of papers reviewed in current issues of a reviewing journal such as *Mathematical Reviews* or *Zentralblatt für Mathematik.*

(vi) Paradox and/or inconsistency

Historians point out two great "crises" during the early period of Greek mathematics, caused by the discovery of incommensurables and the paradoxes of Zeno. These "crises" apparently spurred activity for their clarifications, which was fully achieved by the astronomer and mathematician Eudoxus.

The so-called *mathematical continuum,* or the *continuum of real numbers,* has traditionally been a source of paradox and of attack from philosophers as a breeding ground of inconsistencies. From it the Cantorian theory of the infinite emerged, along with the discovery of new paradoxes and inconsistences (e.g. the Russell antinomy, the set of all sets). However, far from these matters serving as a deterrent to further research,[17] they resulted generally in a challenge to the mathematical world. The stress to clear up the inconsistencies was especially strong and resulted in successful attack by means of formal axiomatics.

---

[17]Some mathematicians did indeed give up research in mathematical foundations, such as Frege, for instance.

Although the professional philosophers apparently resist the solutions that the mathematical community have arrived at with respect to the paradoxes and inconsistencies of the infinite, the majority of mathematicians, especially the platonic-minded, accept them. Consequently the existence of inconsistency cannot be considered an important stress today. However, paradox continues to be a stimulant to research. Topology, especially, has benefited greatly through the paradoxes that its fundamental characteristics engender. Analysts may recall the effects of the discovery by Weierstrass[18] of the real continuous function having no derivative at any point — a purely paradoxical, not contradictory, phenomenon. Mathematicians love the surprise element in their work, and paradoxes serve as a fertile source.

3. General remarks

That what we have called *hereditary stress,* with the six components we have described, is not the whole story regarding the forces generating mathematical progress has already been made clear. Diffusion, generalization, abstraction, etc., not only act individually but often combine; and hereditary stress frequently acts in collaboration with them. But in addition to these cultural forces, we have to recognize the existence of psychological factors.

Among the most prominent psychological factors is that of competition. Even among students of both the pre-college and college level, instructors encourage competition for the solution of problems as a pedagogical device. But among professionals themselves, competition develops in the struggle for promotion in their own departmental groups. We still recognize, however, that such psychological factors are the result of cultural, mainly environmental, stress. The existence of cultures in which competition for higher status is frowned upon is well known to anthropologists. In our culture, however, competition is encouraged — even considered "natural" — and we create prizes and honors — what the sociologist R. K. Merton calls the "Rewards System" — to encourage competition.[19] Journals are instituted for the quick publication of results.

---

[18]Although Bolzano had discovered such a function earlier, its real character was not appreciated (even by its discoverer) at the time.
[19]Merton, 1973; Cole and Cole, 1973.

"Invisible colleges," consisting of scholars sharing a common field of interest, also permit the quick transmission of mimeographed and other inexpensive forms of duplication.[20]

Not all mathematicians approve this situation, and point to the example of the Nobel prizes which exclude mathematics and tend to place the natural sciences on a plane of higher status than mathematics. However, mathematicians have instituted their own international award, the Fields Medals; but these are still known only among the mathematical community, and do not have the popular recognition accorded the recipients of the Nobel prizes.

A related phenomenon is the offering of prizes for the solution of special problems — a device whose effectiveness is, of course, subject to inflation.[21]

We take no sides in these matters, but acknowledge their existence as environmental cultural factors that influence the growth of mathematics. We also recognize the existence of the individual who does mathematics purely for the love of it — "mathematics for its own sake." A favorite example of the latter is the Indian mathematician Ramanujan (1887–1920).[22] The study of the relationship of such individuals to their cultures can be fascinating when sufficient facts about their early lives are known.[23]

Such cases can be recognized and distinguished from those due to internal factors such as hereditary stress, which usually exercises its influence on the professional who is already absorbed in mathematical creation. The psychological aspect of the growth of mathematics is generally a process of reaction to the mathematical culture, and creative synthesis on the part of the individual reacting to this culture.[24]

---

[20]Whether this constitutes "publication" is a moot question, but it certainly furnishes valid evidence of priority of results.

[21]The classic case is the offer, in 1909, of 100,000 marks by a German mathematician, P. Wolfskel, for a published solution of the famous Fermat Problem. As the problem has never been solved, the prize evaporated in the great inflation which followed later in Germany.

[22]For a description of his life, see the essays by Aiyar, Rao and Hardy in Ramanujan, 1962.

[23]In Ramanujan's case, it seems that native ability together with the finding of a book (see Carr 1970) of problems were at the basis of his future accomplishments. One wonders how often a potentially "good" mathematician is deflected by a love of cross-word puzzles and other similar recreations!

[24]For the psychological aspect of mathematical creation, see the classical Poincaré, 1976 and Hadamard, 1949 already cited above.

CHAPTER V

Consolidation: Force and Process

> It is a mark of the great mathematician to have taken a number of separate theories, fragmentary, intricate and tortuous, and by a profound perception of the true bearing and weight of their methods to have welded them into a single whole, clear, luminous and simple.
>
> N. Wiener

Introduction. In Chapter III, during our discussion of important historical episodes, we devoted a section (III-5) to some of the great consolidations that had occurred in mathematical history. In the present chapter, we shall discuss consolidation as a force or process at greater length; and because consolidation is not characteristic of mathematics alone, but of most aspects of culture, we shall divide the chapter into two parts. In the first part we shall discuss it from a general theoretical standpoint; in this way we hope that by keeping it as free from technicalities as possible, we shall make it possible for readers unfamiliar with the more advanced parts of modern mathematics to comprehend the great significance of the notion. The operation of consolidation in mathematics is emphasized in the second part.

Whether consolidation is termed a *force* or a *process* is immaterial from the standpoint of the present chapter. In many of its aspects, it operates like a force; in others, more like an evolutionary process. In biological evolution, is natural selection a "force" or a "process?" Some authors term it a "mechanism" (e.g. Simpson, 1952: 158). However, the choice of nomenclature is not of great importance. The term "consolidation" was chosen since it expresses the operation involved more aptly than the word "combination," for instance, or "fusion."

The term "fusion" can be used, but preferably only in cases where the consolidation results in loss of identity of one of the elements involved. In Wilder, 1953, I used the term "fusion" in several instances where I would now prefer "consolidation;" e.g. in the case of analytic geometry where neither geometry nor algebra (analysis) loses identity. In brief, consolidation, as we shall use the term, is more general than fusion.

Part I

General Theory

Consolidation is one of the most common and outstanding forces involved in cultural evolution. Although recognized in numerous special cases, its universal occurrence seems not to have been adequately emphasized by historians and writers on science. Cultures themselves are undoubtedly the result of consolidation, and the process permeates not only cultural evolution, but biological evolution as well. In the form of what might be termed a "tool," it constitutes an important element of human behavior, especially in the process of invention.

Although we have discussed consolidation to some extent in both EMC and in III-5, it merits a discussion in greater depth, especially in regard to its character and mode of operation, as well as it's importance as an aid to researchers in the sciences. Most of us who engage in scientific research take for granted the logical and other principles by which we are guided; we usually have picked them up from our teachers and the mentors of our first research. And while there seems little use to make a special study of logic for the purposes of the average researcher, the same does not, I believe, hold for consolidation. Although we may use consolidation as a tool intuitively, perhaps automatically, it appears that its possibilities render advisable a constant, conscious awareness of its potential.

As we have observed before (in EMC_1: 168, for instance), as mathematics evolves, there inevitably emerge in the special fields concepts which exhibit similarities that to the perceptive mathematician will suggest new structures encompassing them all. In this way the individual mathematician effects consolidations which, projected into the mathematical culture, contribute to its growth. It is a process with which every beginning research worker must become acquainted as a valuable tool, especially since, as modern mathematics is developing so many and diverse theories, the opportunities for consolidation become more numerous. *Specialization inevitably spawns consolidation.*

In addition to these aspects of its use, consolidation can form an important principle for the historiographer; the recognition of consolidation as a historical phenomenon helps to systematize and categorize certain historical processes. Whenever diffusion occurs, for example, consolidation usually takes place.

Definition. Definitions are always dangerous insofar as they may fail in either accuracy or generality. We have already formulated a definition of consolidation in III-5; in substance it is repeated here:

Consolidation is the uniting of two or more concepts, methods or entities $C_1, C_2, C_3, \ldots$ *to form a structure having greater potential than any of the individual* C_i.

Throughout biology, for instance, consolidations of cells occur to form structures that have properties quite unattainable by the individual cells. In the reproductive process, the interchange and shuffling of genes results in a consolidation that determines the genetic structure of the new individual. Indeed, although biologists seem not to have emphasized it, consolidation appears to be a partner of natural selection quite on a par with the latter in importance.

Chemical mixtures and compounds furnish another example. Aspirin has properties possessed by none of its component molecules of carbon, hydrogen and oxygen. The latter example, incidentally, points up the fact that the manner in which the consolidation is made — in chemistry the bonding character — governs the type of properties achieved by the consolidation. A similar observation can be made regarding mechanical invention.

Among students of the history of mechanical invention, it is commonly recognized that the inventive process is usually one of combining previous inventions to achieve forms which are designed to accomplish new objectives or improve previous inventions. Gilfillan, 1971, stated "invention is a novel combination of old ideas." Classical examples are the invention of the steamboat, which is a combination of boat and steam engine, and the airplane, which combines planes with engines. But such inventions are not merely simple combinations; they involve ingenious sub-inventions as well as careful study of the manner in which the component parts are combined. In the examples just mentioned, special engines had to be invented and adapted to the special situations; and in each case the final product forms a unit in which the original components have been amalgamated, so to speak. The invention is more appropriately thought of as a consolidation than as just a combination.

In social evolution, the consolidation process is of fundamental importance. Formation of the family unit, organization of bands and tribes, the establishment of stable communities capable of developing agriculture and animal husbandry, are primitive examples. And in modern times, industrial consolidations (where the word "consolidation" has been in common use) as well as political groupings are instances worth noting.

It seems scarcely necessary to continue citing such examples; consolidation permeates biological, physical and social as well as cultural processes; so much so as to warrant enunciating a "law:"

LAW OF CONSOLIDATION. *Whenever greater efficiency and/or potential will result, consolidation will ultimately occur; the consolidated entity will then have attributes that none of the original individual entities had.*[1]

Just what the forces are that effect the consolidation in a given case usually constitutes a problem. The physical forces that effect a given type of chemical bonding, the biological forces that enter a particular type of cellular or genetic consolidation, the social or cultural forces that caused the formation of family and other social units, are more or less problematical. Given the consolidated structure, we can perceive its potential for accomplishing certain objectives. Plausible explanations run the risk, especially in the case of social and cultural consolidations, of being purely anthropomorphic.

What can be asserted, however, is the universal occurrence of consolidation in evolutionary processes; it is this, together with the empirical fact that the products of consolidation result in new and more powerful forms, that justifies asserting the "Law" of Consolidation. In every case, time enters into the process in that the current "state of the art" has a great deal to do with the possibility of the consolidation. Even the artist, who traditionally enjoys great freedom in his creations, is nevertheless limited by the cultural background and environment in which he works. Not only is he restricted by the nature of the media available, but he has been nurtured on the works of his predecessors. Beethoven's

[1]In essence, this "law" was first enunciated in EMC: § 6.4, #3. In its present form it occurs in Wilder, 1969: 896. It must be emphasized that consolidation, as an observable phenomenon, is not to be identified with "emergent evolution," as expounded by C. Lloyd Morgan and others. The latter concept had an element of mysticism not assumed here; here consolidation is presumably the result of natural causes.

symphonies, while truly original and outstanding, clearly express a natural evolution from the work of his great predecessors. The natural scientist is restricted in a similar way, but also by experimental data to which his theories must conform, not to mention the state of the technology available for experiments. In particular, the mathematician, while seemingly enjoying as much freedom as the artist, can carry on only from the point which his predecessors have reached.[2] The consolidation which he effects can be built only from the conceptual materials available to him.

1a.   Consolidation as a social or cultural phenomenon

As a social or cultural phenomenon, the consolidation process can be considered from two points of view: from a cultural point of view, or as an individual or conscious process. This is not a true separation, since the motivation leading to an act of consolidation by an individual is usually cultural (e.g. the invention of the steamboat or the automobile). Nevertheless, it is convenient to make the separation for explanatory purposes.

The cultural aspect is represented by the broad consolidation of concepts and fields effected usually over long periods, during which individuals make contributions from time to time, and possibly concluding with the work of one or of more individuals working almost simultaneously. The steamboat, automobile and airplane all furnish excellent examples. Each was the the result of cultural forces motivating, bit by bit, the invention of the component parts and their consolidation. In mathematics, the evolution of the decimal place value system using Hindu–Arabic numerals and Babylonian place value, culminating in Stevin's *La Dismé*, is a good example. In recent years, the tendency in mathematics toward consolidation of concepts and fields has been accelerating.

The individual aspect is, generally speaking, a modern phenomenon. In the opinion of Kroeber, 1948: 352, presumably speaking of mechanical invention: "Deliberately planned or sought invention ... is nearly lacking in most of the history of civilization. It began timidly to come up in Europe around 1300 or 1400, increased in the 1600's, but did not become systematic and important until the nineteenth century. ... In fact, it would have been extremely difficult to plan much invention until both theoretical

---

[2]The case of the mathematician is discussed in greater detail in Wilder, 1950: 264–265.

science and practical technology reached a development, about the seventeenth century, such as had never before been attained." In mathematics, similarly, consolidation of fields would have been difficult until specialization resulted in the variety of fields that have evolved in modern times. However, it did occur in ancient times, as in the case of the consolidation of Babylonian numerical astronomy and Greek geometric astromony by Hipparchus and Ptolemy (see Price, 1961: Chap. 1).

To the individual, consolidation is essentially a *tool,* a device that he uses for purposes of invention; to him, it is purposeful. When he finds that he cannot solve a problem, he looks to other sources than that in which he has been working to see if he can find the means of achieving a solution. As exemplified by the history of the evolution of the airplane, individual acts of consolidation are made in response to cultural stress. In mathematics, we find this "tool" aspect of consolidation appearing most prominently in the atmosphere of "freedom to invent," traditionally attributed to the discovery of the non-euclidean geometries. The present-day mathematician, for example, does not hesitate to borrow concepts from what are considered different mathematical fields. In fact, this type of consolidation can be credited with furnishing tools for the solution of outstanding problems which previously had defied solution. Whereas formerly a mathematician working in a given field ordinarily contented himself with using only the methods characteristic of the field, and may never even have considered that tools and concepts from outside that field might be consolidated with those of his own field of work, today it has become commonplace to look for ideas and suggestions from other fields.

In this way, for instance, such famous problems as the duplication of the cube, trisection of the angle and quadrature of the circle were solved — in the first two cases by using algebraic techniques, in the third by the techniques of analysis. Consolidation of geometric concepts with algebraic and analytic concepts accomplished what classical modes could not do alone.

Ib. Effects of diffusion

Diffusion in the evolution of culture is usually accomplished either through conquest (as in the Christianizing of the native Americans by the Spanish) or by voluntary absorption (as in the adoption of the horse by the

Plains Indians). In each of the parenthetical cases mentioned, consolidation of elements from each of the cultures involved occurred; the native Americans consolidated elements of their own religions with Christianity — the Pueblo Indian, for example, pays homage to his traditional gods as well as hearing Mass in the Catholic chapel. And the Plains Indians consolidated elements of their traditional modes of life with the horse for purposes of hunting and warfare, producing thereby a new culture — a "Plains culture." These two examples point up the fact, however, that consolidation is not always effected smoothly. Christianizing the native Americans was not accomplished without considerable cultural resistance on their part. On the other hand, consolidation of "horse-ways" with traditional modes was generally a "voluntary" process.

One will not expect to find scientific, and especially mathematical, instances of consolidation effected by means of military or political force, although such instances have occurred. When African states and/or tribes were taken over by Europeans, the latter forced their numeration systems on their victims, who consolidated them with their own. But when the Akkadians conquered ancient Sumer, it was apparently the conquerors who consolidated the Sumerian script with their own numeration systems.

"Voluntary" diffusion occurs when it is advantageous to the recipient culture to consolidate aspects of the foreign culture with its own. In mathematics, one of the most recent episodes involving diffusion occurred during the period before and after World War II, when many West European mathematicians emigrated to the United States and consolidated their ideas, often on an individual basis, with those of the American mathematical culture. This was a subtle process, since mathematics had, by this time, taken on much of the character of a world-wide culture. Nevertheless, it occurred, and resulted in the United States taking a leading position in mathematics at the time.

It must be emphasized that in the acquiring of new elements by a culture C_1, the "receiving" culture, from a culture C_2, consolidation inevitably takes place, since C_1 must adapt and modify its own modes in order to absorb the new elements. Indeed, it is difficult to imagine any sort of diffusion from one culture to another being effected without consolidation. The horse was a totally new element in the Plains cultures, but in order to consolidate "horse-ways" with their own ways, the Indians sacrificed their dog-drawn travois and most of the special modes connected with it had to be modified in order to adapt the horse to their culture. Similarly, in West European culture the consolidation of the place-value decimal symbols with the prevailing modes of accounting and other types of record-keeping involved giving up Ionian numerals (which persisted into the 16th century), the abacus, and other types of numeration in vogue.

Historians have frequently remarked on the way in which crossroads of commerce have been centers of advancement in both arts and sciences. The reason for this is the opportunities which such communities offered for the diffusion and consolidation of customs and ideas, which in turn resulted in that greater power that consolidations possess over the elements consolidated.

Part II

The consolidation process in mathematics

IIa. Examples

As an aid in studying the manner in which consolidation has been effected in mathematics, we make a listing of fifteen historically important cases:

1. Adoption of old Sumerian terms for "multiplied by," "find the reciprocal of," etc., by the Akkadians for use as special mathematical symbols.
2. Extension of the Babylonian place-value system to represent fractions; a consolidation of numerical systems.
3. Consolidation of Ionian numerals and Babylonian place value, as in Ptolemy's *Almagest* (cf. III-5).
4. Consolidation of elements from Indian mensuration, Mesopotamian solution of equations and Greek geometric algebra to form Arabic algebra.
5. Consolidation of number with line to form the foundation of early analysis.
6. Consolidation of the concepts of projection as used in art and map-making, with the concepts of Euclidean geometry to form projective geometry.
7. Consolidation of algebra with logic by Boole and others to form mathematical logic.

8. Consolidation of algebra and geometry to form analytic geometry (cf. III-5).
9. Use of algebra and analysis in the investigations of number theory.
10. Consolidation of features of various mathematical structures to form abstract group theory.
11. Functional analysis: Consolidation of the notion of function with that of point in an abstract space.
12. Consolidation of point set topology with combinatorial topology to form the beginnings of algebraic topology.
13. Consolidation of mathematical logic and set theory to obtain solutions of basic problems in set theory.
14. Consolidation of mathematical theories with physical theories to form mathematical physics (a special case of consolidations found in applied mathematics).
15. Consolidation of the infinite with the finite in analysis (e.g. summation of series) and topology (e.g. compactness).

The above is by no means an exhaustive list, of course. Moreover, it is not a listing of the fifteen "most important" consolidations in mathematical history. For example, how can one compare any of the above with the fundamental theorem of the calculus in which differentiation and integration were consolidated; or the consolidation of the complex number with point in the plane giving a foundation for complex analysis similar to the consolidation of real number with line (#5 above)? However, they will serve as examples of how consolidation occurs. More explicitly, they will serve to show how the various forces of cultural evolution enter into the process of consolidation. For usually consolidation does not occur as an isolated force, but is prepared for, accompanied by, or motivated by other forces.

In Part I of this chapter, we have remarked on the frequency with which consolidation occurs during the process of diffusion. How often has this been the case in the above instances of consolidation? Certainly in #1, #3 and #4 the influence of diffusion is clear. Case #1 is an instance of consolidation of symbols between the Akkadian and Sumerian cultures. It appears to have been an especially fortuitous development, since arithmetic is an area in which complete adherence to the natural language is a retardant to operational techniques. Although certain ʻreas of mathematics, particularly those dependent chiefly on logical reasoʻ ʻg (as in synthetic geometry), can progress without special ideographic symbols (although figures — pictorial symbols — are usually employed), arithmetic, like algebra, could progress only when special symbols were introduced. Hence the importance of the consolidation in #1.

Case #3 played an essential role during the consolidation of Babylonian arithmetical astronomy and Greek geometric astronomy — as remarked earlier — a consolidation deemed by Price to be the key to the emergence of Western science (Price, 1961: Chap. 1). Ptolemy's use of the Babylonian tables expressed in the sexagesimal place value system was facilitated by his substitution of the Ionian numerals for the cumbersome Babylonian ciphers; the modern practice which involves the use of degrees, minutes and seconds was, of course, a repetition of Ptolemy's practice, although it is rapidly being replaced by decimal fractions.[3] Clearly stress of a symbolic nature must have been operative in case #3; its presence in case #1 is not so apparent, although still a possibility.

In case #4, where the effects of diffusion are obvious, lie the beginnings of that almost miraculous phenomenon which led to the preservation of fundamental Greek and Oriental mathematical concepts and their ultimate diffusion to Western Europe. (For further details see Boyer, 1968: 254–257). Throughout the general Arabian assumption of ideas from early Greek manuscripts and the later Western European adsorption of Arabic and Greek mathematical works, a virtually continuous process of consolidation went on.

Diffusion is not, however, the only process leading to consolidation in mathematics. Other processes or forces among those important in mathematical evolution lead to consolidations (see EMC: Chap. 4); one of the most outstanding of these is the hereditary stress discussed in Chapter IV. So far as consolidation is concerned, it seems fair to conclude that both hereditary stress and symbolic stress were involved in case #2 (extension of the Babylonian place value system to represent fractions). Manipulation of fractions was presumably an onerous task (as in the use of the Egyptian

---

[3]That is, decimal parts of a degree. The practice of using minutes and seconds and then *decimal* parts of a second was probably a step in the evolution of the consistent use of decimals, although in itself a curious mixture of sexagesimal and decimal systems. Of this, Neugebauer remarks (Neugebauer, 1957: 17): "It is interesting to see that it took about 2000 years of migration (diffusion) of astronomical knowledge from Mesopotamia via Greeks, Hindus and Arabs to arrive at a truly absurd numerical system."

unit fractions, for instance). In Babylonia, the fraction $\frac{1}{2}$, $\frac{1}{3}$, $\frac{2}{3}$ and $\frac{5}{6}$ were evidently in use (cf. Neugebauer, 1957: 26). Unfortunately, little seems to be known regarding the origins of the Babylonian representation of fractions by the place-value system; it seems to have been already known to the Sumerians. So although we can logically argue that hereditary stress, especially *symbolic stress*, was instrumental in the consolidation of place-value notation and fractions, we do not *know* this as a historical fact.

Approaching the modern era case #5 furnishes an example of consolidation clearly forced by hereditary stress. From the time of the Greeks, mathematicians had associated numbers with lines in the sense that a number could be associated uniquely with a line having its length (in terms of a preassigned unit) equal to that number — what the Greeks called a "magnitude". As we have observed before, the Greeks were forced (hereditary stress) to this conception by the problem of handling irrationals, which they were unable to represent in their numerals. Similarly, the early European mathematicians could prove that a polynomial which is negative for a value $x = a$ and positive for a value $x = b$, must be zero at some value of x between a and b simply because its value had to "pass through" zero in proceeding from a to b — at least in the cases where they thought any proof was necessary.[4] It was not until the period of the "arithmetization of analysis," during the latter half of the 19th century, that the isomorphism of real numbers and the geometric line of Euclid was established ("Cantor axiom"). The importance of this, frequently referred to as part of the "freeing of analysis from geometry," was rather that it not only justified the earlier dependence, for both theory and proof, of analysis upon geometry, but the subsequent continued reliance (especially as an aid to the intuition) upon geometric interpretations in real and complex function theory. Moreover, the geometric intuition thereby established was undoubtedly a prime factor in the introduction of abstract space techniques in both functional analysis and set-theoretic topology (see below).

Case #6 was a consolidation of theoretical method borrowed[5] from one area — that of perspective and projection in art and map-making — by

[4]A famous "proof" of this was given by Bolzano in 1817; this constituted recognition of the traditional dependence of analysis on geometry, in that it tried to base the proof on analysis alone.

[5]This was apparently a case of what Elaine Koppelman called "transplantation;" see Koppelman, 1975.

pure geometry to form projective geometry. Cultural stress and generalization were involved; cultural stress because it was cultural needs that first led Desargues, the first of the projective geometers,[6] to apply projective methods to problems of the architect and engineer; and generalization in that he saw in projection "a general procedure for proving theorems about all conics once they were proved for the circle" (Kline, 1972: 300). To be sure, the fundamental work done by Desargues and his followers in the 17th century was destined to virtually disappear by the end of the century, not to be taken up again until over a century later by Poncelet and others.[7]

Case #7 furnishes a most interesting example because of the long period that elapsed between the inception of the fields involved (logic, algebra) and their eventual consolidation. This was due to the long period of evolution that algebra had to undergo before the consolidation could become effective. We recall, by way of contrast, the consolidation of logic and geometry during the Greek era, already noted in III-5 and III-6. Historical details of this are unfortunately hazy, but clearly logic was transplanted to geometry, whereas in the case of mathematical logic, algebra was transplanted to logic. And while the latter is attributed to George Boole, it must not be assumed that Boole created mathematical logic *de novo*. The concept of a mathematical logic was not new, and Boole had predecessors. For a discussion of these, reference may be made to Beth, 1959: Chap. 3. It is interesting that in this case, according to Beth (*loc. cit.*, 55), "A fruitful reform of logic could only be carried out with substantial contributions from the side of pure mathematics. Unfortunately, however, for a long period mathematics was unable to provide the kind of help which logic needed." In other words, before the kind of consolidation which produced mathematical logic could be effected, *abstract* algebra had to be created. It is clear, too, from Beth's description of the evolution of mathematical logic (sometimes called symbolic logic), that the chief cultural force at work in the consolidation was hereditary stress. This acted largely in the form of symbolic stress — exemplified in Leibniz's search for a "calculus of reasoning" — present in the mathematical culture for several centuries. The forces of abstraction and generalization were most certainly also involved in the consolidation, as reading of any good history will testify.

---

[6]The term "projective geometry" was not used in the 17th century, however.

[7]An analysis of the reasons for this phenomenon will be presented in Chapter VI.

Case #8 is the classic example of consolidation, and we have already discussed it in III-5. The basic work in algebra of Vièta, which preceded that of the 17th-century algebraists, was directed toward the application of algebra to geometry, and this endeavour was continued by his successors of the next century, who included both Descartes and Fermat, the "inventors" of analytic geometry. Generally speaking, when a theory is introduced independently by more than one individual (as analytic geometry was by Descartes and Fermat), it is safe to infer that hereditary stress has been instrumental; according to the common expression, the new theory has been "in the air." In this case, we have in addition the fact that the properties of curves, and especially of conics, had become quite important in the 17th century through their use in astronomy, optics and military science, so that evidently environmental (cultural) stress played a part.

In case #9, we are reminded of the solution of the classical problems of geometry (angle trisection, duplication of the cube, squaring of the circle) mentioned in Part I, in that we again encounter the solution of problems originating in an ancient field by appeal to modern techniques of algebra and analysis. Many of the most interesting theorems of number theory, such as the Prime Number Theorem concerning the number of primes less than a given natural number, cannot even by understood without a knowledge of elementary analysis. And contrary-wise, many of the results of number theory are of importance not only for algebra and analysis but for other more modern fields of mathematics such as mathematical logic and topology.[8] The influence of hereditary stress, abstraction and generalization in these instances is abundantly clear.

---

[8]It is gratifying to note that the eminent French mathematician, Jean Dieudonné, has apparently become interested in the general processes characteristic of mathematical evolution. Thus, in a recent paper (Dieudonné, 1975) he has discussed the operation of "fusion," a process (as he uses it) seemingly closely related to, if not identical with, consolidation. He states, "My central thesis . . . is that progress in mathematics results, most of the time, through the imaginative fusion of two or more apparently different topics," citing therewith cases #8 above and the complex analysis referred to above in connection with case #5. He also gives examples of the manner in which analysis can be used in number theory (sections (a) and (b), p. 539, *loc. cit.*). It is not clear, however, whether Dieudonné's approach is confined to the phenomenal aspects of the fusion process, or as part of a cultural approach; I infer the former. I note that I formerly used the term "fusion" synonomously with "consolidation" not only in Wilder, 1953, but in a few instances in EMC.

Case #10, already discussed in connection with generalization in III-7, exhibits a special form of consolidation that is becoming more common in modern mathematics. When certain aspects (such as operations) of diverse theories exhibit a common pattern, then these aspects may be consolidated in a single theory descriptive of the pattern. In the case of group theory, both abstraction and generalization had produced similar theories in the domains of algebra and geometry, and properties recognizable therein furnished the elements for the consolidation defining an abstract group.

It should be recognized that hereditary stress plays a part in all such consolidations of this type, in that the need for singling out special properties from different theories as the basis of a new theory becomes increasingly evident. After the new theory is developed in its own right, it becomes available for the purpose of applications to both old and new branches of mathematics (such as rings, fields, etc., in the case of the theory of groups).

In Case #11, we meet with a phenomenon of the latter half of the 19th century and early 20th century, one in which some of the most characteristic features of modern mathematics become discernible. Here we may have borrowings ("transplantings," diffusion) by analysis from other fields of mathematics, particularly from algebra and topology. Algebraization of analysis started early — during the latter half of the 19th century — and the habit of looking at whole classes of functions, instead of at the behavior of individual functions, led naturally to borrowing from topology for the concept of *function spaces*. Abstract spaces were introduced by Fréchet in his general analysis during the early part of the 20th century, and, later, Banach spaces, Hilbert spaces, etc., appeared. The course of development in the field is too extended and complex to recount here, but fortunately it is adequately presented in publications to which the reader is referred.[9] In this general development one can discern the influence of such forces as hereditary stress, diffusion, abstraction, generalization, as well as sub-consolidations affecting the general consolidation of analysis, algebra and topology. Not to be overlooked, too, in this process, is the way in which effective symbolization eases the treatment of abstract theories.

[9]For example, the beautifully written little book of A. F. Monna, *Functional Analysis in Historical Perspective*, (Monna 1973); or *Encyclopedia Britannica*, 15th ed., 1974, *Macropaedia* vol. 1, pp. 757–772.

Case #12 affords an excellent example of consolidation due to hereditary stress and followed by extensive generalization and abstraction. Until about the 1920s, the field of topology had developed along two lines, one usually termed "combinatorial" and having a finite character making it susceptible to classical algebraic tools, the other termed "continuous" or "set-theoretic" and having an infinite character necessitating the use of set theory (cf. III-2).

In a work entitled *Die Entwickelung der Lehre von den Punktmannig-faltigkeiten* (Schoenflies, 1908) published as a report to the German mathematical society in 1908, A. Schoenflies had attacked problems in plane topology using set-theoretic methods, complemented by properties of polygons. Since he wished to study the form (*Gestalt*) of sets of points in the plane, it seemed logical to him to call upon the already known forms of Euclidean geometry (especially polygons) to approximate the configurations formed by sets of points — a transplantation of Euclidean plane geometry to effect a consolidation with topology. Moreover, he expressed his belief that his investigations could be extended to 3-space and higher dimensions, but that to do so a knowledge of connectivity numbers of surfaces would be required. Such numbers had been introduced by earlier investigators, particularly Riemann, Betti and Poincaré, and formed a major topic in combinatorial topology. By the 1920s, generalizations of these in the homology theory of general spaces reached a point where it was possible to complete the consolidation between set-theoretic and combinatorial topology.[10]

Case #13: During our discussion in I-4 of the ineptness of the "tree" representation of mathematics, we mentioned the manner in which mathematical logic has contributed to the solution of problems in set theory. Such is the case, for instance, with regard to the solution of long outstanding problems such as Cantor's "continuum" conjecture," "Souslin's problem," and various problems of analysis, algebra and topology. In terms of vectors, the symbolic logic and set theory vectors have united their potentials (although retaining their individual identities) to furnish solutions of such problems.

[10]Since this is not a history, we omit details, such as the work of L. E. J. Brouwer in 1910–1913 (which contained the germ of Vietoris' (1927) homology theory) and the fundamental extensions of Alexandroff and Čech to the homology theory of general spaces. For a summary of the details involved in the consolidation described above, the reader may consult the "Historical remarks" in I, § 6 of Wilder, 1949; also see Wilder, 1932.

A typical case is that of the Axiom of Choice (AC) which may be stated thus: For every infinite collection C of disjoint non-empty sets S_i, there exists a set X which has as its elements one and only one element of each S_i. This is a very useful — and controversial — axiom of set theory, but so natural in essence that it was unconsciously used (even by Cantor) until it was discovered. As soon as recognized axioms for set theory (e.g. the Zermelo–Fraenkel — ZF — axioms) were proposed, the problem arose regarding the independence of AC from the ZF axioms.

A well-recognized procedure in symbolic logic for proving independence, based on the use of models, was used (Gödel and Cohen) to show that AC is independent; specifically, AC added to ZF forms a consistent system (if ZF itself does), and a denial of AC added to ZF is also consistent, in that models exist for both of these cases. The so-called continuum hypothesis was handled in the same way. Moreover, a surprising number of set-theoretic statements have been shown to be unprovable from any "reasonable" set of assumptions for set theory. As a consequence one can have many different set theories according to which axioms one chooses to add to such systems as the ZF system.

This is all very unsatisfactory, however, to platonists, who believe that a conjecture such as AC must be true or false, and that showing it to be independent of the commonly accepted axioms of set theory only shows that not enough axioms for set theory have been "discovered" yet.

More precise discussions of these matters may be found in Monk, 1970, Wilder, 1965, and Rudin, 1975.

Case #14 is the classical case of consolidation in applied mathematics. In its results it has worked in two directions: to suggest new mathematical theories as well as new physical theories. Much good mathematics has resulted from the cultural stress imposed on mathematics from physics; and mathematical physics in general has responded continuously to its own hereditary stress by seizing upon mathematical theories to extend and elucidate physical phenomena. Applied mathematics is a fruitful field for examples of a similar nature.

Case #15 has been selected because of its unusual character. The concept of 'the infinite has been the subject of centuries of debate among mathematicians and philosophers of mathematics. Even problems concerning the elementary counting numbers 1, 2, 3, ... (the "natural" numbers) commonly run into questions regarding the infinitude of the class of these numbers. One school of thought insists that these numbers form only a "potential" infinite — i.-e. a class always subject to extension but never actually infinite. The set-theoretic school speaks of "the set of all natural numbers," regarding this as an infinite class. We are not concerned here with the merits of the two philosophies; rather we are interested in how the infinite classes of modern mathematics have been consolidated with an admittedly only finite-capable mathematical framework. For example, for the set of natural numbers, the method of mathematical induction was invented, by which a property known to be true of finite sub-classes of the set of natural numbers can be extended to all the elements of the class. Moreover, for the extension (Cantor) of the natural numbers to transfinite ordinals, transfinite induction and other devices have been invented. In topology, similar devices appear; e.g. extension of properties of finite classes may be extended to properties of a whole class by such principles as the finite intersection property of collections of closed sets (implying the non-vacuous intersection of all the sets in the collection, for instance). For the "infinitely small" the classical "epsilonics" were invented. The obvious motivating force in all these examples was hereditary stress, characteristically revealed in Cantor's *mea culpa:* "I have been logically forced ... to this point of view [acceptance of the completed infinite], almost against my will since it is in opposition to traditions which I value. ...." Again, of course, generalization and abstraction figured prominently in the process.

IIb.   Cultural lag and cultural resistance in the consolidation process

It may seem strange, in view of the great accomplishments due to consolidation, that not infrequently consolidations have been actively resisted (the mathematical analogue of the resistance to new inventions that result in workers' loss of jobs?). After the creation of analytic geometry and the general acceptance of special symbolic modes in the exploration of geometry, strong cultural resistance developed to the traditional synthetic type of proof employed in geometry, and formed part of the complex of factors that resulted in the rejection, by 17th-century mathematics, of the work of Desargues and his followers. Later, however, particularly in the 19th century, there arose on the part of certain eminent mathematicians a strong preference for "synthetic" methods, and on the part of others for an "analytic" approach. Arguments over the respective merits of the two methods ensued, becoming at times acrimonious. "The rivalry between analysts and (synthetic) geometers grew so bitter that Steiner, who was a pure (synthetic) geometer, threatened to quit writing for Crelle's *Journal für Mathematik* if Crelle continued to publish the analytic papers of Plücker" (cf. Kline, 1972: 836).

An analogue of the above occurred in the early 20th century in the resistance evinced by some topologists against the consolidation of algebra and topology, in which some set-theoretic topologists and combinatorial topologists were opposed. One combinatorial topologist complained[11] of the necessity for having to acquaint himself with the sizeable body of literature which had accumulated on the study of abstract spaces, continuous curves, etc., stemming from the set-theoretic school. On the other hand, many set-theoretic topologists rejected the idea of supplementing their tools by algebraic techniques.

The type of cultural resistance exemplified in each of the preceding two cases was no doubt motivated by a complex of causes. One cause clearly evident in the writings of participants was that to the synthetic school (or set-theoretic school), algebraic or analytic methods conceal under a symbolic framework the underlying actual geometric meanings of a proof. In a similar way, Carnot (1753–1823) wanted "to free geometry from the hieroglyphics of analysis" (Kline, 1972: 835). Corresponding to this, the analytic school (or combinatorial or algebraic school) maintained that the brevity and elegance of their methods were worth the price paid in skirting geometric details. It could have been argued, too, that the use of analytic methods in geometry, although a step further away from the "reality" embodied in the geometry, partakes of that greater power which abstraction affords. In some cases there was undoubtedly a reluctance to learn new methods when one's own methodology seemed adequate to the problems in which he was interested. This is presumably the case with the work of many present-day set-theoretic topologists whose concern with general spaces seems adequately served by their methodology. The fact that consolidation has occurred between fields does not necessarily mean that the fields themselves have ceased to operate with their traditional methodology. Synthetic geometry, for instance, continues to survive.

[11]Personal communication.

IIc. Analysis

Although all of the forces enumerated in EMC participate to some extent in the process of consolidation evidently the two most prominent are *hereditary stress* and *diffusion*.

In the case of hereditary stress, two of its components, viz. *capacity* and *conceptual stress* (IV-2(i), (iv)), are the forms most commonly found operative here. When the capacity of a field has been largely exhausted through prolonged mathematical research, then the field is faced either with loss of vitality or renewal through an infusion of ideas from other fields which will increase its capacity. If such ideas exist, then the Law of Consolidation predicts that they will ultimately be found, borrowed and consolidated with the field. Even when a field's capacity has not run out, the conceptual stress created in certain of its aspects may promote borrowing for the consolidation of concepts. The component *challenge* (IV-2(iii)) of hereditary stress is also a not infrequent promoter of consolidation, in that problems of a special nature may arise in a field and invite borrowing from other fields; this is a phenomenon of modern mathematics that is being observed more and more.

The operation of diffusion, which in its ancient manifestations was usually of the inter-cultural type, is today observed more often as occurring between fields. It is true that there still exists a national character in mathematics, although not of the exclusive variety found in ancient times when communication between nations may have been virtually non-existent. This was evidenced to some extent in the consolidation resulting from the immigration to the U.S. preceding World War II, commented upon above. Yet mathematics is essentially a single culture today, so that the diffusion that is most commonly found in case of consolidation is of the inter-fields variety.

Observation of the varying characteristics of the above fifteen cases suggest classification of consolidation according to the kinds of entities consolidated. For example, the consolidation of *fields,* as in case #8, would be one type of consolidation. Unfortunately, there would be a certain amount of arbitrariness encountered, in that the question of what constitutes a "field" would inevitably enter. Another difficulty might be that what may seem at first to be a consolidation of fields turns out more logically to be a consolidation of *methods*.

For example, Case #12, which at first glance appears to have been a consolidation of fields, and was generally considered so at the time, turns out to have been really a consolidation of methods — the set-theoretic method and the combinatorial method.[12] Again, case #2 could be considered either as a consolidation of symbols, or as a consolidation of systems — number systems.

For these reasons, a classification of types of consolidations according to the entities involved appears not to be a promising or very useful endeavor.

Part III
Concluding remarks

In concluding this chapter, we make some general observations concerning the operation of consolidation.

From the definition given in the Introduction, it follows trivially that consolidation requires more than one entity C_i for its effectuation. This reminds us that the opposite force, *diversification* (EMC: 4.1) is a necessary and contributing factor to consolidation. Even the "accidental" type of consolidation resulting from the merger of elements from different cultures could not have occurred without the previous diversification of cultures. In case #1, diversification of language — Sumerian and Akkadian — was a necessary preliminary. In the modern era, the diversification resulting from *specialization* is a most common contributory factor. As a specialty grows, generating new concepts and methods, the analogies with other specialties and their methods may lead naturally to either consolidation of patterns (as in group or category theory) or consolidation of fields. A good share of the uses of mathematics in the other sciences may be attributed to the recognition of mathematical concepts which can be consolidated with physical hypotheses to provide more general theories.

As in the case of inventions, most great advances in science, and especially in mathematics, have resulted from consolidations. Sometimes it is necessary to probe deeply into historical backgrounds to see where consolidation occurred, and it may happen that it is only one of the factors involved. For example, the axiomatic method and its employment of logical deduction, originating in Greek geometry, was apparently a consolidation of philosophical concepts with mathematical concepts,

---

[12]For argument supporting this, see Wilder, 1932, Part I. (The word "unified" used in this article was clearly synonymous with our current use of the word "consolidated.")

forced by the operation of hereditary stress. We have already commented above on the great advances made in astonomy due to the consolidation of geometric and arithmetic theories. The great advance in analysis beginning in the 17th century was facilitated by the consolidation of algebra and geometry. The birth of mathematical logic was accomplished by the consolidation of abstract algebra and logic. Even set theory was originally conceived (Cantor) by a consolidation of number theoretic concepts and the concept of a collection.

Also, the effect on the advance of mathematics and, of course, science in general, of the formation of academies beginning in the 17th century should not be forgotten. Ever since that time, the tendency has been to form organizations, large and small, for the communication of ideas. Today, in mathematics, we have the International Mathematical Union combining all principal national mathematical societies. And on a smaller level, there occurs continual formation of groups oriented both regionally and subject-wise, for the bringing together of mathematicians interested in special problems or methods.

It is difficult to escape the conclusion that consolidation is one of the most important forces in the development of mathematics as a cultural system. Indeed, it appears to be a natural phenomenon that is involved in essentially all movements to attain greater effectiveness and efficiency in achieving man's goals, whether they be control of or adaptation to nature, or the attainment of intellectual objectives.

CHAPTER VI

The Exceptional Individual; Singularities in the Evolution of Mathematics

> ... mathematical discoveries, like the spring-time violets in the woods, have their season which no human effort can retard or hasten.
>
> E. T. Bell

1. General remarks. Mendel, Bolzano, Desargues

In Chapter III we have mentioned a number of historical episodes, illustrative of the manner in which the course of mathematical history has been influenced by certain cultural forces. We have reserved until the present chapter what we shall call the "before-his-time" or "before-its-time" phenomenon; i.e. cases where new concepts appear but are not developed until later, or where individual vectors of the mathematical cultural system have advanced, only to lack the momentum to achieve full recognition. Since new concepts are usually synthesized by individuals, we shall be investigating the reasons why certain individuals made the initial conception as well as why their ideas did not take hold. Such occurrences may aptly be called "singularities" in the history of science.

It is well recognized in scientific research that study of singularities, i.e. of deviations from the norm, is likely to yield valuable information about the "normal" as well as to suggest new concepts. Examples are numerous and are to be found in all fields. The "before-his-time" singularities are identified by their having introduced new theories or techniques which the contemporary related culture ignores or gradually forgets, but which at some later time are either seized upon, or re-created, and developed into mature sub-systems of a field or into new fields of investigation in their own right. To say, in such a case, that the original creator was able, in some mysterious way, to "see" further than his contemporaries, is hardly an explanation of his accomplishment which, in the very nature of the case, is recognized only by hindsight. It is more to the point to ask, *why* did he "see" further, i.e. why did he develop concepts that were, in restrospect, ahead of their time and why did they not "catch on?"

One who is psychologically oriented might be inclined to undertake an analysis of the concerned individual's mental processes in order to seek an answer; but this would usually have to rely on inferences from questionable or inadequate historical data concerning the man's family, upbringing, etc., which could end up with unsatisfactory, possibly mystical, conclusions concerning the man's "genius," or "prophetic vision." While not disregarding these matters, it would seem to be much more fruitful to study the nature of his relationship with his culture and his field of interest, and to seek therein the circumstances which may have triggered his undertaking the task that merited his being considered by later historians as "ahead of his time." We can accept that he was undoubtedly an unusual person, perhaps a genius, but without some motivation and something to base his ideas upon, his creative faculty could not have progressed far.

The case of Mendel is classic. His work on peas was published in a somewhat obscure journal, to be sure, but it was noted in the *Royal Society Catalogue of Scientific Papers,* and it was cited many times by others; and Mendel's discussion of his work in correspondence with Nägeli bears further witness that his work was not overlooked. It was simply *not understood,* at least in its implications for a science of heredity. The details are readily available; a thorough investigation has been made of the motivation for Mendel's investigations (see, for instance, Iltis, 1966 and Zirkle, 1951) and the recency of the case makes it easy to ascertain why his results were not given the prominence that they deserved until long after his death. To quote Conway Zirkle (*ibid.*), "We are ... justified in emphasizing a remarkable coincidence. Before Mendel, the component parts of Mendelism had been discovered separately, some by the plant hybridizers and some by the bee breeders. Very few biologists were cognizant of the data which had been acquired in both of these fields. Mendelism was the creation of an investigator who both hybridized plants and bred bees." In short, Mendel probably made a remarkable consolidation of concepts in preparing his experiments and arriving at his theory of heredity. Furthermore, although the reasons for his theory being ignored at first are easy to find, he was not so far ahead of his time as to preclude his work being discovered in time to serve as a basis for the modern theory of genetics.

Not all progenitors of a theory fare so well. Mathematicians are quite familiar with both concepts and theorems which have been assigned names of individuals who were re-discoverers rather than originators. An analogous situation holds even for broad theories. A case analogous to that of Mendel, perhaps, is that of Bolzano (1781–1848), a Bohemian priest, much of whose work on the foundations of real analysis escaped attention until after credit for discovery had been accorded to others. Furthermore, the topological nature of his thinking, adumbrative of modern dimension theory, is coming to be recognized (see Johnson, 1977).

A more interesting phenomenon in mathematics, particularly because of the long interval that elapsed between the discovery and its recognition as the basis of a new field of mathematics, is that of the French military engineer and architect Desargues. In 1639 he published a work which could well have formed a foundation for projective geometry, a subject unknown at that time. But it was unappreciated in much the same manner as Mendel's work, and was ultimately lost and essentially forgotten for approximately *two centuries*! His was a clear case, in all aspects, of being ahead of his time, and quite rewarding of study.

One can expect that when a mathematical theory disappears, it will be principally due to cultural factors, either intrinsic, environmental, or both. The creative individual is always available to carry on a theory which has sufficient potential; if it dies out, the reasons should be sought elsewhere than in the lack of investigators. The 17th century was remarkable for its host of mathematical talent; the names Descartes, Fermat, Pascal, Newton, Leibniz, among others, come to mind. Desargues' case has been studied by historians, of course, at first by Poudra (1864), and later (and more authoritatively) by Taton (1951a). Also, special articles have been devoted to it, such as Taton (1951c, 1960), Swinden (1950) and Court (1954). Consequently the main details regarding the man and his work are available, and some well agreed upon reasons have been given for rejection of his work.

Our plan here will be to study the matter from a culturological point of view, going into the cultural factors that prompted Desargues' work in the first place, that contributed to its disappearance, and, for the sake of comparison, the reasons for its re-creation about two centuries later by mathematicians unfamiliar with his work. The whole makes, we feel, an instructive analysis of a clear-cut case of being ahead of one's time, both as regards the cultural milieu before and after the event, and its later break-through in a more hospitable mathematical culture.[1]

2. Historical background of Desargues' work

We shall recall briefly the most pertinent historical details. Since the name "projective geometry" was not introduced until the 19th century, and for the sake of brevity, we shall use the symbol "PG17" to denote the 17th-century work of Desargues and his followers in what later became known by the more familiar name "projective geometry."

The case of PG17 is all the more remarkable to one who is familiar with the usual patterns of mathematical evolution, since it appeared as a result of cultural conditions which one would naturally expect might produce it — although as will become apparent, not under conditions that would foster it. Renaissance painters of the 15th and early 16th centuries had initiated the revolt against the stilted and unrealistic type of art that dominated the medieval period, and turned to "realism." Fortunately these men, unlike the modern specialists whom we call "painters" or "artists," were fundamentally and of necessity engineers skilled in the construction of buildings, fortifications and bridges, in addition to being painters and mathematicians. Realism in painting required a study of perspective in order that three-dimensional illusion might be presented on a two-dimensional canvas. Both Leone Battista Alberti (1404–1472) and Albrecht Dürer (1471–1528) studied and wrote on perspective, and Leonardo da Vinci (1452–1519) and Piero della Francesca (*ca.* 1416–1492) were prominent in the insistence on a geometric foundation for art (see Kline, 1953, Chap. 10).

The basis of perspective lay in considering the eyes as a point and the rays of light between this point and the object to be painted as a cone. The canvas on which the painting was to be done was thought of as a screen held between the eyes and the object, thus playing the role of a section of the cone. Practical problems such as how to transfer such a section to the actual canvas had to be solved. One theoretical problem, destined to prove of great importance to mathematics later, was proposed by Alberti: What

---

[1]My interest in the case of Desargues was originally derived from a suggestion regarding the "inhospitality" of 17th-century mathematics to projective geometry, contained in C. B. Boyer's review (*Science,* Feb. 21, 1969) of EMC.

properties of an object are preserved in a section; or, more generally, what properties of the object do different sections have in common? This work on perspective led to a scientific art of painting, and some of the best mathematicians were to be found among its practitioners.

During the second half of the 16th century, demands of commerce led to the employment of projection in map-making, although this represented a change in type of application, and new kinds of projection were involved (Boyer, 1968: 327–329).

2a. Girard Desargues and "PG17"

Although knowledge of his personal life is meager, apparently Girard Desargues (1591–1661) was skilled in both architecture and engineering, and had also acquired an interest in geometry of the classic type. He became particularly interested in the techniques employed by such craftsmen as stone-cutters and dial-makers, whose laborious use of geometric rules led him to use his superior knowledge of geometry to simplify and codify the problems in perspective encountered by them. In 1636 he published a small text on perspective embodying his ideas and in 1640 a brief essay on perspective for sketching designs in architecture and for sundials. According to Taton (1951c, p. 10), this work shows a clear comprehension of the basic principles of descriptive geometry which lie at the foundation of these techniques.[2]

He seems, however, to have run into professional jealousy, and to have become embroiled in a series of quarrels with J. Curabelle, Beaugrand (the King's secretary) and others. His sensitivity to these affairs led him from 1644 on to refuse to publish any more.[3]

Desargues' interests were not, however, confined to the practical arts. He was evidently familiar with the newly translated works of Apollonius on conics, as well as with Pappus' related works. During the 1620s he had become a member of the Paris circle of Mersenne, which met weekly to discuss mathematical and philosophical questions. He became a close

[2]Sarton, 1950: 300–301 quotes Professor Pierre Huard of Indochina to the effect that "one of the early books introduced into Japan was Desargues' treatise on perspective (*Manière Universelle*) in the Dutch translation by Jan Baron, 1964)." A list of works by Desargues is given in Swinden, 1950, and Taton, 1951a gives the text of the *Brouillon Projet* (see below).

[3]See Taton, 1951a: 98. Swinden, 1950, gives a detailed account of Desargues' quarrels with his enemies.

friend of René Descartes, who also had a strong interest in geometry. But while Descartes' approach to geometry was through algebra, Desargues was attracted to the purely synthetic type of argument. And while his knowledge of geometry had led him to applications, his proficiency in the theory of perspective led him to consider what its methods could do for geometry. The result, in a nutshell, was a kind of consolidation of the techniques of perspective with the study of the properties of conics. He came to see that the derivation of Apollonius' results could be greatly simplified through studying conics as projections of the circle and determining the properties of the former that are carried over invariantly from the circle.[4]

Desargues' classic work, the one which might have been rightly termed the beginnings of projective geometry, and which we term PG17, was a treatment of conics having the title *Brouillon projet d'une atteinte aux événemens des recontres d'un cone avec un plan*. Published in 1639, it was a highly original and brilliant work, certainly on a par with any work of his contemporaries. But the quality of a work is not the sole determinant of its acceptance in the macrocosm of learning, and it proved to be, in the words of Boyer, "one of the most unsuccessful great books ever produced" (Boyer, 1968: 393). Only around 50 copies were printed and distributed to a selected group. All of these copies disappeared until one was found about 1950 in the Bibliothèque National, Paris, containing notes and errata by Desargues himself. However, fortunately a manuscript copy made by Philippe de La Hire was discovered earlier — in 1845 — so that the later geometers of the 19th century were not entirely deprived of a knowledge of Desargues and his work. But in the meantime, so completely did Desargues' PG17 and the ideas in it vanish that, according to N.A. Court, "Writing less than a century after Desargues' death, the renowned (historian) Montucla ... scarcely knows Desargues' name. ... All he can tell is that Desargues was a friend of Descartes and that he wrote an essay on conics which at the time was appreciated by some mathematicians of repute" (Court, 1954: 7).[5]

[4]For a more detailed discussion of the evolution of Desargues' ideas, see Taton, 1951a: 95–98.
[5]Some of Desargues' critics, in particular Curabelle, mentioned a work of Desargues entitled *Lecons de ténèbres*. Apparently no trace of this has been found. Swinden, 1950: 257 n suggests this title was an alternative for the *Brouillon Projet*.

It should be noted that geometry was not in disfavor at the time; much of the work in algebra shortly before and concurrently was directed at applications to geometry. Moreover, Desargues' work was done in a center of mathematical activity containing such men as Descartes and Fermat, who admired it. It is true that Desargues' book with its ponderous title was couched in a strange terminology much of which was borrowed from botany; but it presented a new and powerful method for solving problems about conics.

At first it seemed that the young Blaise Pascal (1623–1662) would continue the work done by Desargues on PG17. But after obtaining a few brilliant results, Pascal turned to other fields of interest (Kline, 1972: 295–298). Another likely prospect, the above-mentioned Philippe de La Hire (1640–1718)[6] published in 1685 a work, *Sectiones Conicae,* devoted to PG17. He seems to have proved most of Apollonius' theorems on conics, using projection and section, and "tried to show how projective methods were superior to those of Apollonius and the new analytic methods of Descartes and Fermat which had already been created" (Kline, 1972: 295). A lesser figure, Abraham Bosse (1611–1678), also wrote a book entitled *Manière universelle de M. Desargues, pour pratiquer la perspective,* whose chief claim to fame seems to be that we owe it to our knowledge of the famous so-called "Desargues Theorem," which later turned out to be so fundamental.[7] Taton, in his book on Monge (Taton, 1951b: 67), mentions also a work by Le Poivre, entitled *Traité des sections du cylindre et du cône considérées dans le solide et dans le plan,* Paris, 1704, in which conics were studied from the projective point of view.[8]

There were also works on pure geometry (more or less in the tradition of Desargues and Pascal), worth noting from a historical point of view, during the 18th century. One in particular, the famous Scotch mathematician Maclaurin, whose name is attached to a well-known series expansion in calculus textbooks, did some noteworthy work in pure geometry. Whether he knew Desargues' work, either directly or indirectly, seems undecided. In

[6]Not to be confused with his father, Laurent de La Hire, the painter, who was among the enthusiastic disciples of Desargues' work in perspective.

[7]See, for instance, D. Hilbert's *Foundations of Geometry,* LaSalle, Illinois, Open Court, 1971, Chapter V.

[8]Taton's remarks in this connection (1951b: 67–68) are recommended reading.

the Introduction of his famous *Traité des Propriétés Projectives des Figures* (Poncelet, 1865), Poncelet states that in his geometric researchers (*ca.* 1720 to 1750), Maclaurin took up ("reprenant") the work of La Hire having, no doubt, no knowledge of the writing of Desargues and Pascal. He recalls, however, that La Hire mentions the works of Desargues in the preface to his *Sections coniques.*

Others who worked in pure geometry (as opposed to analytic geometry) during the same period (18th century) were the Englishmen Braikenridge and R. Simon, and the Swiss-German mathematician Lambert, whose *Traité de Perspective* (1774) used perspective methods to establish many propositions.

However, by the end of the 17th century, interest in PG17 had essentially disappeared, while the analytic geometry of Descartes and Fermat had prospered and been incorporated into the work of Newton and Leibniz on the calculus. Not until the 19th century would PG17 be "reincarnated."

3. Why was PG17 not developed into a field?

It is hardly possible for us to place ourselves, in our imaginations, in the social, and particularly mathematical, environment of 17th-century France. (We cannot enter even a contemporary "primitive" culture and appreciate or understand thoroughly its ways, so immersed are we in the customs and ideals of late 20th-century Western culture.) We know that during the 16th century, applications of algebra were made to geometry, and that in this fact perhaps lies the kernel of the 17th-century application of algebra in the form of analytic geometry. Of more importance to us however, is this indication that geometry was in the highest repute.

Why, then, was PG17 allowed to die? As we have already stated, Desargues' work was fully as brilliant as anything to be found in the contemporary analysis. True, his book was limited to about fifty copies, sent to selected readers who presumably could understand and appreciate it — and no doubt did (Descartes, for example). We have today "invisible colleges" — small groups of specialists who "publish" their work in mimeographed form for distribution among themselves; yet their work continues to survive and, in many cases, results in the establishing of new journals. It is difficult to believe that the paucity of copies of Desargues' book was much of a contributing factor to the demise of PG17. Mathematics at the time was not institutionalized, and it was quite customary to publish one's work in a few copies and circulate them among possibly interested parties.

Also, as we pointed out above, Desargues used a strange terminology, borrowed in many instances from botany. But mathematics has customarily borrowed its terms from the natural language; consider "continuous," "limit," "function," "category," "inverse," "mapping," etc. And botanical terms can be found among the terms of graph theory, which seems to have flourished despite this. Moreover, apparently La Hire and Bosse found no difficulty interpreting Desargues' work despite its unusual terminology. And R. Taton remarks (1951a: 97) that the terminology of the *Brouillon Projet* does not create major difficulties for the attentive reader.

The major causes of the demise of PG17 must be sought elsewhere. It is not enough to say that mathematicians of the day did not like it; those who read it and understood it seem to have admired and praised it, although they may have exercised their own talents not in PG17 but elsewhere such as in the new analysis fathered by analytic geometry. And as stated above, it was not that synthetic geometry was not important, for it was; analysis could not have got along without it until the 19th-century foundations of analysis were laid in the arithmetization of the real continuum. Nevertheless, the majority of mathematicians favored the new analytic geometry over PG17 as a field for research. Why? The answer can be found in both (1) the state of the mathematical environment at the time, and in (2) the internal nature of PG17.

3a. The mathematical environment of the 17th century

Although mathematics could hardly be considered as a cultural system in the early 17th century, it was beginning to develop more under the influence of internal forces than formerly. Although no professional mathematical associations had yet developed, there did exist scientific academies and, in France, the informal but influential group centered about Mersenne. The latter seemed to serve as a focal point for dissemination of scientific and especially mathematical results. Nevertheless, the conditions prevailing in the French culture of the 17th century should definitely be of prime consideration in any study of the scientific element of the time.

But there was another external element seemingly not heretofore emphasized in Desargues' case. Every historian is aware of the fact that a mathematical theory cannot be effectively developed until the conceptual framework necessary to nurture it has been prepared. For example, the logic of the Greeks remained a rhetorical exercise until abstract algebra had advanced to the point where Boole could achieve an algebraic formalization of it, thereby setting the stage for modern mathematical logic. For the evolution of formal logic it was the *conceptual* background, not the formal manipulative aspect of algebra, which was needed, and especially the concept of algebra as a "free" science, not anchored to the science of arithmetic.

Similarly, one of the major factors needed for the survival and extension of PG17 was the concept of "free" geometries, i.e. geometries not restricted to the Euclidean framework which still dominated geometric thinking in the 17th century. Not until the 19th century was it possible even to *think* of a geometry founded on the concept of transformation and invariant (finally culminating in Klein's *Programm*). Poncelet's recognition of projective geometry as a distinct field of geometry, in his classic of 1822 (the first edition of Poncelet, 1865), and especially such a work as the *Geometrie der Lage* of von Staudt, published in 1847, in which both measure and congruence were deleted from the definition of cross-ratio, would have been unthinkable in the atmosphere of the 17th century. It could hardly have been expected, then, that PG17 would appear as anything other than an extension of classical geometry.

Another contributing factor was the above-mentioned lack of institutionalization of mathematics in the 17th century. The "mathematicians" of the time, such as Descartes, Fermat and Pascal, were not teachers in universities where "captive" audiences might afford a means of sparking interest in one's ideas. The situation was otherwise in the late 18th century, when the great geometer Monge was teaching at the École Polytechnique (for the founding of which he was largely responsible). A great teacher, he numbered among his students such men as Lazare Carnot, Poncelet, Michel Chasles and others. These men helped assure the continued evolution of Monge's geometric ideas. But Desargues, not a teacher at any institution, acquired no such following (except for the famous Pascal, who turned to other interests after proving the famous theorem named after him).[9] Without disciples to continue its growth, a mathematical system cannot survive as a living mathematical entity.

Of course, institutionalization is not a *necessary* condition for the survival of great concepts, as witness the work of the contemporaries of Desargues and other well-known "loners" of mathematical history such as Abel and Galois, but along with other cultural and societal factors, it can certainly be a great aid.

3b. The internal nature of PG17

First, let us look at PG17 from the standpoint of the ideas introduced in Chapter IV. Of course, PG17 did not achieve the status of what we normally call a "field." The word "field" has now acquired a status character exemplified in such terms as "the field of algebra," "the field of topology." One would not ordinarily use the word to designate a special topic such as determinants (although this has been done). PG17 was essentially the work of one man, Girard Desargues; his fellows added little to his work. Nevertheless, as a new methodological theory, it was evidently known and approved by such eminents as Fermat, who expressed admiration for it, and Pascal, who contributed to it, and it was well enough known to create public opposition. It was by no means a piece of mathematical trivia, meriting only aloofness from the world of scholarship. Indeed, when 19th-century geometers learned of it, they gave it high praise — so much so that historians seem generally to regard it as the genesis of projective geometry. But, of course, this is hindsight.

The *capacity* of PG17, i.e. the quantity and intrinsic interest of the results which the basic theory of PG17 was capable of yielding, can be best studied in comparison to the capacity of its prototype, the projective geometry of the 19th century. In the latter case, it became clear that a new geometry was evolving, one more general than the classical Euclidean geometry, based on transformation and invariant with no need of a metric, and hence having greater capacity. But the history of PG17 shows that both Desargues and his 17th-century contemporaries considered that the basis of PG17 was Euclidean geometry, of which it was deemed to be an extension. (Compare

---

[9]Pascal called Desargues "one of the great intellects of this time and amongst those most expert in mathematics."

Kline, 1972: 300, 3rd paragraph.) That it revealed new properties and introduced more powerful and general methods of proof did not gainsay its Euclidean nature and hence its having no more capacity than the traditional geometry of Euclid. Apparently the creative mathematicians of the time concluded that to work in PG17 would be to indulge only in a further "gilding of the lily" and to offer little future. The example of PG17 brings out clearly that the capacity of a system such as PG17 is partly a function of the time as well as of the general state of mathematics itself.

Similar remarks hold for the *significance* of a system. From the viewpoint of hindsight, we can see that PG17 constituted a very significant development for mathematics; but this was by no means apparent in the 17th century, and it was its significance at that time which was crucial. As Taton observed (1951b: 68), "Ainsi, pour la seconde fois, le rajeunissement de l'edifice géométrique tenté par un homme de génie, à la fois technicien et géomètre, echouait, et la géométrie, ignorant volontairement les méthodes nouvelles que la technique et l'art suggéraient, se figea dans une attitude de respect quasi superstitieux des méthodes héritées de L'Antiquité."

Similar conclusions follow from a consideration of the *challenge* of PG17 — i.e. the emergence of problems whose solution would require an ingenuity and/or methodology which distinguishes them from problems whose solutions are of a routine character. The initial challenge of PG17 came from problems in perspective encountered by the Renaissance scholars, such as the problem of Alberti cited above. These problems presented a challenge requiring both an ingenuity and new methodology, which Desargues was able to supply. But PG17 itself was rather a diversion into pure mathematics whose Euclidean nature apparently presented little challenge. Moreover, the problems of a qualitative character which it did present held little attraction for an environment in which quantitative aspects of mathematics were becoming increasingly important. The phenomenon which we perceive in number theory, which has continued to survive down through the centuries chiefly because of its never-ending supply of challenging problems (despite the changing character of the other mathematical disciplines), was not to occur for PG17.

That PG17 lacked *status* (the esteem in which the field was held among mathematicians) should already be clear from previous remarks, as well as from a consideration of its "worker ratio," i.e. the ratio of the number of mathematicians contributing to it, to the number of creative workers in mathematics at the time. The large majority of those who were working in mathematics was absorbed in algebra and analysis, fields whose significance for the developing sciences was soon to become increasingly evident; and the appeal of the algebraic method for studying curves was proving an attractive novelty which the limited method of projection could not offer. So lacking in status did PG17 become that its methods were termed by some as "dangerous and unsound" according to Boyer, 1968: 394.

It is interesting, however, that in contrast to the above deficiencies, PG17 was not lacking in *conceptual stress,* i.e. the stress created by the need for new conceptual materials in order to furnish a logical basis for explaining phenomena. For instance, the concept of *line at infinity,* adumbrated by Kepler's point at infinity, was effected by conceptual stress. Indeed, this aspect may be considered to have been the most hopeful for PG17. Had it only been strong enough to bring to the fore the basic concepts of transformation and invariant, rather than settling for the study of conics within the framework of the Euclidean tradition, projective geometry might conceivably have evolved earlier than it was destined to do. There apparently existed a willingness to adopt new conceptual attitudes, without thought for such philosophic concerns as existence or reality; but the impulse in that direction was insufficient. As stated above, the changed mathematical climate of the 19th century was needed to provide the necessary stimulus.

4. Avenues of possible survival

There were still two avenues of possible survival open to PG17 if the pattern taken by other systems in like situations could have been followed. One of these was consolidation with other parts of mathematics. This may be said to have actually happened to the Euclidean geometry of the time through its consolidation with algebra to form analytic geometry. But, except for the consolidation implied by Desargues' ability as both a technician and geometer, consolidation was not to occur in the case of PG17, in the way that its successor, the projective geometry of the 19th century, was to consolidate with algebra and analysis when analytic methods which had not been developed in the 17th century became available. What looked like an attempt by Desargues' follower La Hire to effect some kind of consolidation with algebra in his *Nouveau éléments des sections coniques* in 1679 did not succeed (Boyer, 1968: 404).

The other possibility was the diffusion of new concepts from other mathematical disciplines into PG17. This was to occur in the 19th century in the case of projective geometry, when Cayley, von Staudt and Klein introduced new concepts through algebraic and analytic means, resulting in the recognition of projective geometry as the basic geometry, of which other geometries, such as the Euclidean and non-Euclidean, are sub-fields. But, again, this was not to happen to PG17, simply because the concepts needed from other fields were not in existence at the time.

5. The success of projective geometry in the 19th century

It is revealing to consider why projective geometry made such a successful reappearance in the 19th century. In particular, how did it get started? We recall that PG17 was created by Desargues after he had produced manuals on the use of projection and perspective for the craftsmen of the day, especially for the stone-cutters. Interestingly, the start of projective geometry nearly two centuries later was quite similar. Gaspard Monge (1746–1810), although not himself the originator of projective geometry, became involved in using projective methods for his system of drawing plans for military fortifications; he seems to have invented descriptive geometry in the process (some of which was restricted as military secrets!)[10] Unlike Desargues, he had the advantage of university teaching at the École Normale and the École Polytechnique. It was in this capacity that he inspired such men as Poncelet and Chasles, both of whom took part in the development of projective geometry. We have here, incidentally, an example of the great teacher whose creative talents inspire a formation of a "school," in this case a school of geometry.[11]

Poncelet was the first to develop projective geometry to any great extent, his ideas being a natural outcome of his studies under Monge and with

[10]For the influence of Monge on modern geometry, see Taton, 1951b: §6, 273–276, entitled "Monge, précurseur de la géométrie moderne."

[11]Regarding Monge's abilities as a teacher, Taton, 1951b: 273, says: "Grâce à ses qualités pédagogiques exceptionelles, il sut présenter les éléments de cette nouvelle science [descriptive geometry] avec toute la clarté et l'élégance désirables, attirer sur elle l'attention de nombreux élèves et diriger l'enthousiasme de ceux-ci en leur indiquant les nombreuses voies nouvelles qui pouvaient s'ouvrir à leurs recherches."

Carnot, who had also entered the field. Apparently he had no inkling of the *Brouillet Projet* of Desargues, except through a copy of a letter of Beaugrand (one of Desargues' opponents; see Poncelet, 1865: xxvii), which had been preserved and which Poncelet termed "truly worthy of a snarling critic." Poncelet's work, *Traité des Propriétés Projectives des Figures,* was the result of researches he did in the spring of 1813, while imprisoned in Russia as a soldier in Napoleon's army. It was published in 1822 (2nd edition, 2 vols., 1865, 1866) and is subtitled, *Ouvrage Utile à Ceux qui S'Occupent des Applications de la Géométrie Descriptive et d'Opérations Géométriques sur le Terrain.*

Since we have no intention to review the development of projective geometry in the 19th century, but only to look at its beginnings, we shall not go further than to remark that with the publication of Poncelet's work, the rest is well-known history. In a broad way, the evolution of PG17 and the early evolution of projective geometry are remarkably similar; both Desargues and Monge started their geometric studies in applications of geometry, the former in perspective for craftsmen, the latter in descriptive geometry for architecture and particularly design of fortifications. Both went on into pure mathematics.[12] But according to the tacit rule by which "firsts" are assigned to those who produce the discoveries that directly influence the ultimate developments, one could argue that it is Monge, Poncelet, and the 19th-century geometers who should receive credit for projective geometry, not Desargues, even though no historian would write a history of projective geometry without first describing Desargues' work. This is not to belittle Desargues' genius. The high praise that his work elicited from his 19th-century successors, when the *Brouillet Projet* was discovered, was clearly deserved. His case differs from Mendel's, however, in that the work of the latter was discovered in time for it to form the basis of modern genetics. Desargues' work was discovered too late to provide the basis of projective geometry. His was a unique genius, however. He was not the only one to work on the mathematics of perspective; but he was the only one in his century to go further and to create a new geometry, even though neither he nor his contemporaries could recognize it as such in the then current state of the culture. All the other recognized men of genius in mathematics of that era, especially Descartes and Fermat, were "in tune

[12]When, later in the century, he became familiar with Desargues' work, Poncelet called Desargues "The Monge of his century" (Poncelet, 1865: xxv-xxvi).

with the times," and showed no inclination for synthetic geometry. To be sure, had Desargues never lived, Pascal might be considered as one who *could* have created projective geometry. But (1) he proved his disinterest as soon as he had published his famous essay on conics and other similar isolated works, and (2) history shows that a background of work in perspective, or similar work involving projection, was necessary for the initial steps in projective geometry — and this Pascal did not have. In short, PG17 was the product of a unique creative mind and a special background in gnomonics, stone-cutting, engineering and general perspective, which would most likely not have occurred in the 17th century without the presence of Desargues. On the other hand, from its basis in Euclidean geometry, PG17 could properly be considered only as an ingenious extension of Apollonius' work on conics (which is probably the way in which his mathematical colleagues regarded it), rather than as the *initiator* of projective geometry.

From the standpoint of the cultural evolution of mathematics, it is interesting to speculate on the position of projective geometry in mathematics as a whole. Can one say of it, as in the analogous statement about the calculus and Leibniz–Newton, that it would have inevitably appeared even though Desargues, Monge, Carnot, Brianchon, Poncelet, etc., had never lived?

The situation in the late 18th and early 19th centuries was, of course, far different from that in the 17th century. PG17 is a good example of the dangers inherent in a theory's evolving too soon. The mathematics of the time was simply not ready to apprehend or utilize its capacity and significance, nor, more important, to afford it the concepts and tools for its further development. But by the late 18th century, not only had more extensive work been done on perspective, but the teaching of the rudiments of descriptive geometry had begun in the newly established military and technical schools, and research in the new geometry became more extensive; see, for example, Poncelet, 1865, *Préface de la Première Èdition.* In short, all signs pointed to an extension of the concept of projections from descriptive geometry to projective geometry; a reading of the historical details is quite convincing. Of course, we have the benefit of hindsight, and it is easy to say that "the time for the development of projective geometry had come." Certainly it was trying to break through in the 17th century, as PG17 shows. That it succeeded in doing so in the 19th century was due to the confluence of both environmental stress (e.g. demand for courses in descriptive geometry in military and engineering schools, and growing institutionalizing of mathematics with consequent establishment of a role status for mathematicians and encouragement of research in pure geometry) and internal (hereditary) stresses, especially challenge and conceptual stress, working through the creative mathematical minds of the time.

6. General characteristics of the "before his time" phenomenon

Desargues and Mendel are not, of course, the only "singularities" of their type in the history of science. Even in the time of the Greeks, one may presumably consider Aristarchus, whose heliocentric theory preceded Copernicus by some 1800 years, as an early member of the fraternity. One may with some justice consider Saccheri (1667–1733), who preceded Bolyai, Gauss and Lobachewski by approximately a century, as being ahead of his time in his geometric researches on the independence of the parallel axiom, which were published in 1733. Of course, Saccheri's aim was to *prove* the parallel axiom from the other axioms of Euclid; he had no notion of such a thing as its *independence,* presumably. He was really in the tradition of Omar Khayyam, Nasir al Edin, Wallis and Lambert, all of whom had tried to prove the parallel axiom. However, as eminent an authority as T. L. Heath pointed out in his presentation of Euclid's *Elements* (Heath, 1956: I, 211), Saccheri's work is "of much greater importance than all the earlier attempts to prove Postulate 5 [the parallel axiom] because Saccheri was the first to contemplate the possibility of hypotheses other than that of Euclid, and to work out a number of consequences of these hypotheses. He was, therefore, a true precursor of Legendre and of Lobachewsky, as Beltrami called him (1889) and, it might be added, of Riemann also. For, as Veronese observes (*Fondamenti di geometria,* p. 570), Saccheri obtained a glimpse of the theory of parallels in all its generality, while Legendre, Lobachewsky and G. Bolyai excluded *a priori*, without knowing it, 'the hypothesis of the obtuse angle,' or the Riemann hypothesis." Had the state of his culture been different, he might very well have been able to form the correct conclusion from his work, and hence have been a true precursor.

Charles Babbage (1792–1871), with his computers, has been termed by

Boyer as "an eccentric" who "lived a century before his time" (Boyer, 1968: 671–672). According to the 14th edition of the *Encylopedia Britannica,* Babbage may be called "the originator of the modern automatic computer."[13] E. T. Bell (1945: 92) mentions the discovery in 1936 by the Polish scholar K. Michalski that William of Occam (1270–1349) proposed a three-valued logic. Of C. S. Peirce (1839–1914) it has been commented: "... upon reading his papers, one cannot help but be amazed at how many ideas were expressed, if not fully developed, by him years before their supposed discovery" (Lewis, 1966: 46). H. Grassmann developed what we now call tensor calculus, which was later developed further by C. G. Ricci (1888). But if it had not been for Einstein's use of the tensor calculus in his theory of relativity, Grassmann's work in this regard might well have gone unappreciated.

To facilitate the remainder of this discussion, we shall denote cases of the "before his (its) time phenomenon" by the term *prematurity*; and the individuals concerned by "premat." What characteristics can one detect as likely to be found in a premat and a prematurity? The following appear to be prominent in the cases that have been mentioned.

6a.   The premat as a loner

By this we do not necessarily mean loner in the social sense; one can be active socially while still a loner in his intellectual life. We have already mentioned the late Indian mathematician Ramanujan in the closing remarks of Chapter IV. Whether he could be called a "loner" in his social life is questionable, since his ties to his family and friends in India seem to have been very close. Indeed, the influence of the Indian culture on him was apparently strong. Hardy wrote about him as follows: "it [Ramanujan's work] has not the simplicity and the inevitableness of the very greatest work; it would be greater if it was less strange. ... He would probably have been a greater mathematician if he had been caught and tamed a little in his youth; he would have discovered more that was new, and that, no doubt, of greater

---

[13]According to Dubbey, 1978, Babbage was "far in advance of his time" in his algebraic views, having "anticipated many of the early theories of so-called modern algebra" (p. 81); and his work on the calculus of functions which "contains some very original strategems and devices" had not been accorded a just recognition. His work in pure mathematics was done during the period 1813–1821 after which he spent some fifty years on computers and in other areas.

importance. On the other hand, he would have been less of a Ramanujan, and more of a European professor, and the loss might have been greater than the gains" (*Ramanujan,* 1962: xxxvi).

Until his voluntary exile, Desargues seems to have led a normal social life, but in his ideas about geometry, he was clearly a loner. However, it is probably more often the case that the premat is also a loner socially. Certain clerical professions may be inducive to lack of social contacts with one's scientific colleagues; Mendel, Saccheri and Bolzano are examples. C. S. Peirce spent the last 26 years of his life at his home in Milford, Pennsylvania.

Cutting the ties to one's culture can lead one to the creation of theories that are foreign to the ways of the culture; and in particular, a scientist who works in exile from his contemporaries will likely not interest them with his creations nor be understood by them.

6b. Tendency of the premat to create a vocabulary that repels possible readers

This seems to have been a prominent factor in the cases of Desargues, Grassmann and Peirce. Each of them built up conceptual structures while making few compromises with the terminologies used by their contemporaries. Of course, any new theory will ordinarily require new terms and definitions, but these can usually be related to existing terminological systems in such a way that an abrupt break in the nature of the terminology can be avoided.

6c. The capacity and significance of the new concepts embodied in the prematurity not recognized

This is not necessarily a corollary of 6b, although a strange terminology can contribute to it. Mendel's work did not suffer from poor exposition or strange terminology, and elements of it had been produced by others before him (see Zirkle, 1951). His contemporaries completely missed the potential in his work, nevertheless. In Desargues' case, the curious terminology and the brevity of the exposition did prove an impediment, but lack of recognition of the capacity and significance of his work proved more of a factor in its rejection.

6d. The culture not ready to incorporate and extend the
new concepts embodied in the prematurity

Babbage encountered this difficulty. His first computer, the "difference engine," was subsidized to the extent of nearly a million dollars by the British government. But apparently the machine tools needed for its construction were not in existence at the time, and Babbage spent ten years inventing and manufacturing tools and needed parts, at the end of which time the government withdrew its support.[14] Abandoning the project, Babbage conceived in 1853 the idea of an "analytical engine." This again failed because of lack of financial support and engineering problems. The analogy to the case of Desargues is striking; in the latter, the mathematical tools of a conceptual nature were not in existence, and the cultural need was lacking.

6e. Lack of personal status of the premat in the scientific community

Grassmann suffered from this, since he was a teacher at a gymnasium, and was hardly known even in Germany. Mendel, too, as well as Bolzano, lacked the status necessary for early recognition of their work.

An obvious contribution to lack of personal status, especially in medieval times, was distance from the major center of activity in the related field. Even today, this is a factor, although modern means of communication have lessened its effect.

6f. Insufficient diffusion of the new ideas presented by the prematurity

This is sometimes due to repelling or uninviting terminology noted in 6b or it may be due to the lack of disciples. This is not likely to occur today, of course, with modern close cultural contacts, and in the modern institutionalized world of science, but it is a factor to be noticed in historical cases of prematurity. Even today, however, lack of students to carry on one's work can lead to prematurity (or extinction!)

---

[14]The Swedish government later financed the construction of a modified form of Babbage's machine which was completed in 1853 and sold in 1856 to the Dudley Observatory in Albany, N.Y., where it was used in the construction of tables.

6g. An unusual combination of interests on the part of the premat

In Mendel's case, this was pointed out by Zirkle 1951, and has been mentioned above. In Desargues' case, the combination of technical knowledge involving perspective (stone-cutting, gnomonics, etc.) and an intense interest in geometry, especially in the properties of conic sections, certainly played a role. Babbage furnishes another example of the combination of theory and technical expertise. Such unusual combinations do not necessarily lead to prematurity; indeed, they probably occur more often in the normal evolution of theories. They do seem to be present in a high percentage of the cases of prematurity, however.

7. Comment

The presence of any or all of the above characteristics is not, to be sure, sufficient to produce prematurity. Without the individual motivation and the environmental stress that produces the creative urge in the individual, such singularities would never occur. Indeed, every creative scientist may be thought of as a potential premat, inasmuch as his creations are excursions into the unknown. If he follows an unknown path too far, he runs the danger of finding himself isolated from the community of scholars to which he properly belongs; he then becomes a loner. Typically, a premat is the scientist who creates a conceptual world that his colleagues either ignore or do not recognize as of importance, but which later finds its proper place in the universe of science. The premat may realize that his ideas are not of importance at the time, yet be so attracted to them that he is willing to forego the recognition that ordinarily is accorded among the rewards of scientific activity. The scholar who maintains close contact with the way in which his field is evolving and chooses his problems from among those of current interest, may safely avoid becoming a premat! But, too, he may produce less of what is significant in the future!

CHAPTER VII

"Laws" Governing the Evolution of Mathematics

> ... when we cannot trace natural phenomena to a law ... the very possibility of comprehending such phenomena ceases.
>
> H. L. F. von Helmholtz

> Mathematical science, like all other living things, has its own natural laws of growth.
>
> C. N. Moore

Most cultural systems exhibit patterns of behavior that are frequently described in the form of "laws." The "law" of supply and demand in an open market has been a favorite of economists, and in the definitions of physical systems, "laws" abound. All such "laws" have an ideal character; for example, the law of falling bodies is never exactly exemplified outside the laboratory. Yet the success of a theory, insofar as approximating "reality" is concerned, is totally dependent on the application of such "laws."

In EMC: V-4, we gave a tentative list, ten in number, of "laws" that seem to govern the evolution of mathematical concepts. Subsequently the historian and philosopher of science, M. J. Crowe, gave ten "laws" concerning patterns of change in the history of mathematics (Crowe, 1975 a, b), and in a review of EMC_1 Crowe (1978) gave a discussion and critique of the "laws" stated therein. In a subsequent commentary (Wilder, 1979) we reacted to Professor Crowe's review, thereby hopefully clarifying and extending our ideas. In the present chapter, we again take up the subject in the light of both Professor Crowe's and our own contributions to the subject.[1]

In using the word "law" we do not intend, of course, a statement concerning fixed and immutable patterns to which the evolution of

---

[1]See also Koppelman, 1975, in which six categories of steps are identified in mathematical change.

mathematics must conform; rather we mean to indicate certain forms in which evolution has taken place, but which, like a physical law, may run into exceptions. In short, the laws constitute recognition of the fact that certain patterns tend to repeat themselves, and that prediction of future change and/or events can be made on their basis with a high probability of success.

We shall state these laws one at a time, discussing each before proceeding to the next one.

1. Multiple independent discoveries or solutions of outstanding problems are the rule, not the exception

See Crowe's law 8 (1975a) and our law 4 (EMC). For a discussion see II-1.

An outstanding possible exception to this law occurred only recently, in the solution of the famous Four Color Problem. See Appel and Haken, 1977 (although more than one author was involved, the solution was a joint venture). It is possible, of course, that simpler proofs may be forthcoming, possibly not involving use of computers (as was the case in the solution just referred to), and possibly coming in the form of a multiple, so that we cannot be positive in this matter. In fact, the following law, although related to #1, states a general pattern:

1a. The first proof of an important theorem is usually followed by simpler proofs

The reason for the pattern stated in #1 is no doubt the hereditary stress generated by the significance of the theorem and particularly the challenge presented by the call for simpler proofs.

2. New concepts usually evolve in response to hereditary stress or, alternatively, to the pressure from the host culture as expressed through environmental stress

It is hardly necessary to argue the fact that hereditary stress is a prolific instigator of mathematical invention; it is virtually a corollary of the *continuity* of mathematical evolution. On the one hand, a concept cannot evolve until the groundwork for it has been prepared; Boole could not have invented symbolic logic if the required algebraic concepts had not been developed. On the other hand, a concept will not likely evolve until it is needed for some mathematical purpose, or to fill an environmental need; the Babylonians could hardly have invented infinitary concepts even though their number system was extendible without end.

As for the influence of environmental stress on the evolution of new concepts, we have always before us the counting numbers — the natural numbers — which were patently a result of the necessity for them in their host cultures. That two-way street, mathematics ↔ physics and the invention of new mathematical fields due to war-time exigencies (e.g. operations research), and especially the invention of the electronic computer and its accompanying theory, all testify to the influence of environmental stress on the creation of mathematical concepts.

3. **Once a concept is presented to the mathematical culture, its acceptance will be ultimately determined by the degree of fruitfulness of the concept; it will not be forever rejected because of its origin or because metaphysical or other criteria condemn it to be "unreal"**

By "fruitfulness" we intend not only benefit to mathematics from the problem-solving capacity of the new concept, but its power to open new areas of mathematical research; also the aesthetic satisfaction it may afford to its practitioners. It should be remarked, however, that there can be a lag between the announcement of a new concept and its ultimate "fruitfulness;" this has been amply attested to by "before-their-time" phenomena (Mendel, Desargues, etc.).

The assertion concerning rejection of a concept is included in #3 in recognition of the fact that new concepts have so often met with antagonism and rejection on the grounds of "unreality." The case of the complex numbers (indeed, the *negative* numbers) and the case of the Cantorean theory of the infinite are both illustrative of the assertion.[2] It is unlikely, however, that in today's mathematics a concept would be rejected on grounds of "unreality" — unless, to be sure, there were adopted some kind of constructive mathematics which considered as "unreal" all non-constructive parts of mathematics.

[2]Cf. Crowe's laws 1–3 and his discussion thereof in Crowe, 1975a: 162–163.

4. **The fame, or status of the creator of a new mathematical concept exercises a compelling role in the acceptance of that concept, especially if the new concept breaks with tradition; a similar remark holds for the invention of new terms or symbols[3]**

We have already pointed out, in connection with the discussion of symbolic achievements in III-2, that "the status of the individual proposer in the mathematical community usually has an important effect on whether the symbol becomes accepted by the mathematical community and hence part of the mathematical culture." In addition, we can hardly do better, by way of substantiating this law, than use Crowe's words (*loc. cit.*):

> Compare the reception accorded Hamilton's *Lectures on Quaternions* (1853) with that of Grassmann's *Ausdehnungslehre* (1844). Both are among the classics of mathematics, yet the work of the former author, who was already famous for empirically confirmed results, was greeted with lavish praise in reviews by authors *who had not read his book,* whereas the book of Grassmann, an almost unpublished high school teacher, received but one review (by its author!) and found, before it was used for waste paper in the early 1860's, only a handful of readers. Or consider the fate of Lobachevsky and Bolyai, whose publications remained as unknown as their authors until, thirty years after their publications, some posthumously published letters of the illustrious Gauss led mathematicians to take an interest in non-Euclidean geometry. [Italics mine — R. L. W.]

5. **The continued importance of a concept or theory will depend both on its fruitfulness and on the symbolic mode in which it is expressed. If the latter tends toward obscurity while the concept maintains its fruitfulness, then a more easily manipulated and comprehendible symbolic representation will evolve.**

One might question the validity of this law in the light of the many centuries over which algebraic symbolism evolved. It was a time of individual proposals for symbolic improvements as well as of rejection of cumbersome symbols. Moreover, during the earlier centuries there was obviously little hereditary stress for good symbolic modes — hardly surprising in view of the as yet meager development of mathematics. By the time of Vièta, however, some stress had been built up; this is evidenced by the relatively short time in which the calculus received its own special

[3]Cf. law 6 in Crowe, 1975a: 164.

symbolization. The fruitfulness of the calculus as the basis of classical analysis and its capability in response to environmental stress (mechanics, physics, etc.) led inevitably to the selection of the Leibnizean symbolic representation.

In his somewhat controversial *Development of Mathematics* (Bell, 1945),[4] E. T. Bell states that

> Unless elementary algebra had become "a purely symbolical science" by the end of the sixteenth century, it seems unlikely that analytic geometry, the differential and integral calculus, the theory of probability, the theory of numbers, and dynamics could have taken root and flourished as they did in the seventeenth century. As modern mathematics stems from these creations of Descartes, Newton and Leibniz, Pascal, Fermat, and Galileo, it may not be too much to claim that the perfection of algebraic symbolism was a major contributor to the unprecedented speed with which mathematics developed after the publication of Descartes' geometry in 1637. (*loc. cit.*, p. 123.)

During the 18th and 19th centuries, realization of the importance of good symbolism became widespread. Cayley introduced matrices which, in their subsequent history, provide an example of the suggestiveness and fruitfulness of a good symbol. Boole symbolized logic. Plücker devised his system of homogeneous coordinates for projective space. The tensor calculus of Ricci proved fundamental for Einstein's work in relativity.

6. **If the advancement of a theory comes to depend on the solution of a certain problem, then the conceptual structures of the theory will evolve in such a manner as to permit the eventual solution of the problem. As a rule, the solution will be followed by a flood of new results.**

The solution of the problem may take centuries, as in the case of the problem of the relationship of the parallel axiom to the other axioms of Euclidean geometry, which was followed by the work on the new geometries. A better example, perhaps, is the problem of the solvability of

---

[4]This work, first published in 1940 (2nd edition 1945), contained numerous errors of a technical nature. But its broad sweep, and easy diction interspersed with wit, made it very popular. It is interesting to note that in the same year in which the second edition of Bell's book was published, he began a paper for a memorial volume honoring D. E. Smith with the words, "Not being in any degree a historian, I ... "

equations by algebraic means ("solvability by radicals") of Ruffini, Abel and Galois. The surge of work in the theory of groups and the exploitation of algebraic fields and abstract algebra followed.

The work on *finite* groups, characteristic of the latter part of the 19th and first part of the 20th centuries, is of special interest. After a flurry of activity, the theory of finite groups was finally declared "dead." One of the legacies of the theory was the so-called Burnside conjecture to the effect that every non-Abelian simple group has even order. This conjecture stood as a challenge awaiting proof for more than half a century, and was finally proved true in 1963 (Feit-Thompson, 1963). The result has been a resurgence of life in the once "dead" subject of finite groups.

In analysis, the problem of the real continuum, the analogue of the problem solved by Eudoxus with his theory of proportionality, was solved by the 19th-century mathematicians (Dedekind, Cantor *et al.*) thereby making it possible to forge ahead with new theories of integration (Borel, Lebesgue *et al.*) and other portions of what is now called "classical analysis."

7. **If a mathematical theory will be advanced by, and especially if its advance depends upon, the consolidation of certain concepts, then this consolidation will take place**

The idea expressed in this law is in essence much the same as the contention that such inventions (consolidations) as the automobile and airplane were inevitable. Given the direction in which technology was advancing, these inventions were inevitable.

We have already mentioned in III-5 the consolidation in topology of the set-theoretic and combinatorial (algebraic) methods. Although this was a "logical" consolidation, it actually resulted from the presence of problems whose solutions could only be expedited by the consolidation.

Again, in the case of analytic geometry, the state of algebra and its relation to the rest of mathematics made inevitable its consolidation with geometry. Evidence for this is afforded by the fact that it evolved through a multiple (Dedekind, Fermat).[5] At the time algebra had just undergone a rapid development (especially symbolic) by Vièta and others, a process

---

[5]As we have argued earlier, evolution of a concept by a multiple is virtual proof of the inevitability of its development.

culminating in Descartes' *La Géométrie.*[6] The latter was actually an interplay between geometry and algebra with roots going back to al-Khowarizmi.

The best evidence for law 7 is undoubtedly presented by 20th-century mathematics since World War II, where consolidations are commonplace. Although some specialties appear at first glance to be self-sufficient, closer examination shows important consolidation, especially as to method. In V-1 we have called this an operation of the Law of Consolidation; we repeat it here for the sake of completeness:

7a. (Law of Consolidation). *Wherever greater efficiency and/or potential will result, consolidation will ultimately occur; the consolidated entity will have attributes possessed by none of the entities consolidated*

As remarked in Wilder, 1969: 896, this seems to be a general law of nature.

> Thus, among living entities, it is exemplified by the consolidation of cells to form new living structures having properties unattainable by the elements entering into their makeup. In chemistry, examples abound; aspirin has properties possessed by none of its component molecules of carbon, hydrogen and oxygen. In sociology, we see congregations of individuals forming political or social consolidations which can achieve a wide variety of tasks impossible for the individual. In economics, mergers occur between related industries and, recently, even nonrelated industries. In fact, everywhere in nature and society we observe this trend toward consolidation. The result of the operation of the Law of Consolidation in mathematics is the power to solve problems that have hitherto defied solution. Research in mathematics has become more and more a search for structures and relations, representative of conceptual frames over the whole broad spectrum of mathematics. Structures which in effect consolidate two or more branches of mathematics are likely to be the most effective in their mathematical and scientific fruitfulness.

8. Whenever the advance of mathematical evolution requires the introduction of seemingly absurd or "unreal" concepts, they will be provided by the creation of appropriate and acceptable interpretations

The material already given in II-7 seems adequate discussion of this law. Compare also Crowe 1975: laws 3, 9 as well as the above law 3.

---

[6]The notation in this work, except for the sign of equality, was similar to that used today.

9. At any given time, there exists a cultural intuition shared by members of the mathematical community, which embodies the basic and generally accepted opinions concerning mathematical concepts [Wilder, 1967; Crowe, 1975: law 5]

Before discussing this law, it is necessary to make clear what is meant here by "cultural intuition." We first recall what we stated in I-3 regarding the nature of our mathematical culture: "The bulk of our beliefs, mores, and technology, have congealed into a heterogeneous mass of elements, whose origins are lost in unwritten history ... [Herein] there exists a 'tradition' composed of [the working] logic, mathematical folklore, ... , etc." The "intuition" referred to in the above law is part of the "mathematical folklore" referred to in this citation. It contains the beliefs concerning basic mathematical concepts which are taken for granted by most mathematicians. These beliefs are an accumulation of assumptions garnered from mathematical experience; their existence is taken for granted until they are forced into the open through the evolution of mathematical concepts.[7]

Since many mathematicians might doubt the existence of such an intuition, we shall resort to known historical events which served to bring it into the open; such events usually served also to change the intuition in important respects.

The apparent belief of the early Greek mathematical community, and especially the Pythagoreans, in the commensurability of all geometric magnitudes also constituted an intuition that was finally shattered. The solution of the problem thereby created, embodied in Eudoxus' theory of proportion, seems to have heralded a period of intensive activity in Greek geometry.

For over two millennia, mathematicians (and others who pursued mathematics as a hobby) believed in the absolute truth of mathematics, and particularly that the axioms of Euclid's *Elements,* including the famous axiom of parallels, were *true.* The impact of the eventual realization that the axiom of parallels could be replaced by axioms contradictory to it, thereby producing consistent non-Euclidean systems, led not only to drastic change in the original intuition, but also to the creation of varying

---

[7]For a more extended discussion of mathematical intuition, see Wilder, 1967; also see Bunge, 1962.

philosophies concerning what constitutes "permissible" mathematics. Of perhaps not so great immediate influence, the demonstration by Hamilton and Grassmann of the possibility of non-commutative algebras served to change the intuition that all algebras and numeral operations must be commutative.

By the advent of the 19th century there had been built up an intuition concerning real continuous functions. When one thought of the graph of such a function, it was usually in the form of a curve having no breaks in it. Moreover, although such curves might have "angular" points at which the function had no derivatives, the common intuition was that at "most" points of the curve, the function had derivatives. The breakdown of this intuition is owed to Weierstrass (1861); an earlier example (1830) given by Bolzano was not generally known (cf. law 4), nor did its creator apparently realize its complete character.

Because of its fundamental nature, topology, particularly of the set-theoretic variety, has provided many "cures" for false intuitions. For example, it was long believed that a plane closed curve, i.e. a common boundary of two domains in the plane (such as a circle), could have only two complementary domains, i.e. its complement in the plane must be just two connected open sets, of which it is the common boundary. This intuition, that a closed curve could have only an "inside" and an "outside," may very well have been due, at least partially, to the influence of the Jordan Curve Theorem, which had created wide interest in the mathematical world. The intuition was so strong that it was taken for granted by A. Schoenflies, one of the principle founders of the topology of Euclidean spaces, in his fundamental work *Die Entwickelung der Lehre von den Punktmannigfaltigkeiten* (Schoenflies, 1908).[8] The shattering of this intuition by L. E. J. Brouwer led to a series of fundamental researches in topology by both Brouwer and others.[9] Incidentally, the discovery that a plane closed curve may be the common boundary of more than two (even an infinity) of domains was a multiple, having also been discovered by the Japanese mathematician Wada; see Yoneyama, 1917–1920.

Another example, from analysis, relates to the parametric representation of plane curves and ultimately to the definition, by C. Jordan, of

[8]Other aspects of this work were discussed above in Chap. V, case #12.
[9]This is not to imply that Brouwer's work did not have other stimulants; see obituary by H. Freudenthal and A. Heyting in vol. 2 of Brouwer's *Collected Works* (Brouwer, 1976).

continuous curve in the form $x = f(t)$, $y = g(t)$, where f and g are continuous single-valued real functions over the interval $[0,1]$. There seems no doubt that in giving this definition in his *Cours d'Analyse*, Jordan was giving explicit expression to a prevalent intuition. However, Peano, E. H. Moore and Hilbert all soon gave examples of such "curves" whose graphic expression occupied all points of the unit square plus its interior. These examples of "space-filling" curves were followed by many others (cf. law 1a). More important, however, was the period of activity in research on both continuous curves and plane topology which followed.

Probably one of the most devastating (to the intuition) examples was that of Banach and Tarski (1924). Specifically, they proved that "A solid unit sphere can be decomposed into a finite number of pieces which can be re-assembled to form *two* solid unit spheres." This seems intuitively impossible, and has been used to question the validity of the Axiom of Choice, on which the proof depends. (From this axiom it can be shown that not all three-dimensional point sets have volume (or measure)).

Each field of mathematics develops its own collective intuition, as of course do its individual practitioners. Usually the individual intuitions agree with one another on major issues, coinciding with the collective intuition, but not always. Without these intuitions — collective and individual — research in mathematics would hardly be possible. The individual mathematician usually tries to prove what he feels intuitively to be "true." The one who succeeds in proving a proposition that runs counter to the collective intuition of his speciality creates the most attention in the long run.

10. Diffusion between cultures or fields frequently will result in the emergence of new concepts and accelerated growth of mathematics, always assuming the requisite conceptual level of the receiving entity[10]

As observed in III-1, diffusion may occur geographically (between cultures) or intra-culturally (between fields). In the former case, the diffusion may occur to fill in a cultural gap — a "missionary-type"

---

[10]See law 5 of EMC. The qualifying clause "always assuming ... " answers a criticism made in Crowe, 1978: 104.

phenomenon in which the receiving culture is brought up to a higher level of technical competence. This is illustrated by the diffusion of mathematics from the Arabic culture to Italy and Spain (see III-1), as of course also in the diffusion of elementary mathematical facts from more advanced cultures to primitive cultures. This type of diffusion has been going on for some time between European–American cultures and primitive African cultures,[11] and cannot be expected to result in the creation of new mathematical concepts until the receiving cultures have reached the requisite cultural level.

The case of China presents a highly interesting situation, in that prior to the recent revolution, Chinese mathematicians were already making essential contributions to mathematical progress. Some of the most eminent current mathematicians are of Chinese extraction, although usually connected with American and European institutions. The situation regarding the mathematicians on the Chinese mainland has been clouded by the prevailing cultural environment, but now that the opportunities for diffusion are apparently increasing, one can hope to see important contributions to mathematics from China proper.

Of the other major cultures, India presents one of the most interesting pictures. During the domination by British culture, the diffusion of mathematics to India seems to have been mainly of the "missionary type," probably due to cultural resistance. Beginning with the enrollment of Indian students in English, and later in American and other universities, the capability for mathematical conceptual thinking adumbrated by the ancient Hindu philosophers was soon evidenced.

One of the most prevalent problems in such cases, particularly in modern times when ease of transportation from one area to another is so advanced, is that the "receiving" culture becomes also "giving," in that the students whom they send to other cultures for training tend to remain within the confines of the host culture. This is, of course, due to the desire for cultural contacts — fellow scholars with whom communication is possible, library

[11]We use the word "primitive" only to exclude those cultures (South African and Egyptian, for instance) which have had higher education facilities for some time. While some of this diffusion, particularly between European and African cultures, was directed only toward bringing the receiving (African) culture to a level requisite for trade purposes, today's efforts are also aimed at bringing the receiving cultures to the fullest possible level of attainment. It should prove interesting to see how soon the latter African cultures begin to contribute to mathematical advancement in modern areas of research. See Zaslavsky, 1973.

resources, etc. At times such factors are complemented by political conditions, as in the case of many Chinese scholars who did not wish to return to mainland China because of political conditions.

The most noted and recent case of diffusion of a "non-missionary" type occurred when German mathematicians fled their homeland in terror of the Nazi regime. The resulting diffusion was especially noticeable between German and United States scholars, in that the majority of European scholars (including Polish and Hungarian) fleeing the terror ultimately went to the United States. The resulting interaction was, until World War II at least, a diffusion of ideas and methods between equally knowledgeable domestic and foreign scholars that proved highly productive.

Diffusion from one mathematical field to another is usually to fill a recognized need in the receiving field. There is no question of conceptual level involved in such a case. Whenever one field of mathematics borrows concepts from another, it usually consolidates them with already existing concepts of its own. The amazing advances in the modern field of topology have been to a great extent due to its consolidating topological concepts with concepts from both algebra and analysis. Such diffusions have resulted in permanent changes, although exceptions may occur; for instance, the recent solution of the four-color problem, which drew upon the resources of the computer, will probably not lead to any permanent diffusion from computer theory to topology.

Diffusion between mature mathematical fields frequently are reciprocal; concepts in topology resulting from the borrowing of algebraic concepts has later led to the borrowing of parallel concepts by algebra. Similarly, number theory and analysis have each benefited from ideas of the other; and the classical relations between geometry and analysis have formed a two-way street.

Diffusion aided by consolidation has proved a major force in the evolution of mathematics. Even a casual reading of mathematical history confirms this, and it it has occurred both inter-culturally and intra-culturally. The diffusion of inter-cultural type was prominent in pre-modern times, the most notable being that of the diffusion from Babylonian and Egyptian cultures to the Greek culture, and that of the Arabic to the European cultures. In modern times, due to the essentially universal character of the mathematical culture, the predominant type of diffusion has been intra-cultural — between fields of mathematics. The predominance of this type of diffusion has been one of the chief contributions to the accelerated evolution of mathematics observable during the past century. With diffusion of mathematical ideas to the developing countries, however, one may in the future find that it is in this inter-cultural type of diffusion that the greatest acceleration may be found.

11. **Environmental stresses created by the host culture and various subcultures thereof, such as the scientific, will elicit an observable response from the mathematical subculture. The character of this response may be either an increase in the creation of new mathematical concepts or a decrease in mathematical production, depending upon the nature of the stress**

As a subculture, mathematics constitutes only one vector in the system represented by its host culture, however this system may be conceived. As such, it will be subject to the pressures exerted by other vectors of the system. As an example, consider the military vector in time of war. Throughout its evolution, mathematics has been increasingly called upon during war times for assistance to the military; and with the institutionalization of mathematics as a separate vector, the effects are more easily observable. In particular, during World War II, involvement of mathematicians in the many technical aspects of warfare had a lasting effect on post-war mathematics. This was apparent in such areas as computer technology, which received a substantial boost during the war years. Similar remarks may be made regarding statistics, in which, for instance, the idea of sequential analysis was developed.

In addition to such effects on already established fields of mathematics, new fields were developed due to the war. Operations research, so important to the study of flexible gunnery in the air force, was represented after the war by the establishment of university courses and, in some cases, new departments. Linear programming grew out of wartime experience in staff planning needed by the United States military establishment. The effects of such developments on mathematics proper is debatable. Establishment of "split-off" departments in computer theory, departments of statistics, departments of operations research, etc., as well as incorporation of mathematics into schools of business, engineering and the social sciences, have at least temporarily resulted in the diminution of mathematics departments. Such developments have also apparently resulted in a lessening of research in core mathematics.

The effects of military developments on mathematics during the French Revolution and the Napoleonic era in France form a notable chapter in the history of mathematics. Monge, especially, whose work on descriptive geometry was kept secret by order of the military, became involved in the establishment of the École Polytechnique, whose influence on such mathematicians as Lagrange, Legendre, Carnot, Poncelet and Chasles is a matter of record. This institution was also responsible for the establishment of the first journal devoted exclusively to mathematics, viz. *Journal de l'École Polytechnique.*

One can add, with some reservations regarding historical accuracy, the name of Archimedes, whose work was presumably continually interrupted for the purposes of devising "engines of war."

A different form of environmental stress, mainly economic and occurring during periods of high unemployment, has led students away from the study of core mathematics and into more "applied" university courses such as computing, statistics and actuarial courses. The effects on core mathematics due to the decreasing number of students are much the same as in an industrial field — the fewer workers, the less production. On the other hand, in times of economic depression, when the overall number of students may decrease, the opportunities available for new workers in the area of research in core mathematics decrease with a corresponding decrease in the output of research. This phenomenon appears to be occurring presently (1979) in the United States, and occurred also during the depression of the 1930s.

A new phenomenon occurred, however, with the ending of World War II, viz. the rise of both private and governmental agencies devoted to subsidizing scholarly and scientific research. In the United States the military (first the Office of Naval Research and soon thereafter the Army and Air Force) commenced giving grants and/or fellowships to assist in mathematical research. The government established the National Science Foundation, which provided funds for both individual and institutional projects in mathematical research and mathematical education. These environmental factors have had a marked effect on the progress of mathematics, and the process still continues as of the present (1979). Private foundations, such as the Sloan Foundation, have also been established for the purpose of assisting in a financial way with the expansion of research in mathematics and the sciences.

The natural sciences, with their great dependence on mathematics, both as a tool and as a source of concepts, have always provided a stimulus to the creation of new mathematics. A classic example is the work of Fourier in the theory of heat, in which the types of new functions introduced spurred the mathematical community to new investigations in the theory of functions, with repercussions in the theory of sets. Most of the classical mathematics was created by, or suggested by, environmental needs of workers in science, commerce, architecture, and other pursuits active at the time.

In modern times, however, the operation of hereditary stress is undoubtedly a more important factor in the evolution of new mathematics; so much so, that one can speculate on how the growth of mathematics would be affected by a severe economic collapse. In the latter case, it seems likely that the economic factors discussed above, resulting in lack of employment, would ultimately dominate the situation and result in a virtual cessation of productive research in mathematics. There seems little question, for instance, that the gradual decline of Hellenistic mathematics was due to the prevailing decline in scholarship in general in response to the conditions heralding the approach of the Dark Ages. It seems that the overpowering dominance of the host culture can at times virtually obliterate those vectors — scholarly, artistic, scientific — that may be deemed the least necessary for the survival of the culture.

12. **When great advances or breakthroughs are made in mathematics, then, allowing for time while the mathematical public absorbs their implications, there will usually result new insights into concepts previously only partially understood, as well as new problems to be solved**

The qualification regarding "time" is not so important today, when means of communication are so adequate. We recall, however, that the implications of the Gauss–Lobachewski–Bolyai non-Euclidean geometries were not appreciated until around a quarter of a century after their inception, with the publication of Gauss's papers and Riemann's *Habilitationschrift* (Riemann, 1854). And it is only in hindsight that we can realize the effects of the formulation of the calculus by Leibniz and Newton in the latter part of the 17th century; the burgeoning of work in analysis that followed attests to the impact of this great advance and the legacy of problems to be settled. The new insights which gradually took shape regarding the nature of limit, the notion of function, etc., were extended into the centuries following.

Certainly the importance of the theory of sets, during the latter years of the 19th century, was not realized by the general mathematical public; indeed it was a time of skepticism and rejection. Only gradually was it realized what a tremendous effect this introduction of new concepts was to have on the development of mathematics during the ensuing 20th century.

One can only conjecture the impact of the discovery by the Greeks of the method of creating geometrical properties by logical deduction from a basic set of postulates; or the effect of Eudoxus's theory of proportion. Unfortunately the historical details here are lacking.

13. Discovery of inconsistency or inadequacy in the current conceptual structure of mathematics will result in the creation of remedial concepts (see II-9)

The classic example of this is, of course, the discovery by the Greeks of inconsistencies and the remedying of the situation by Eudoxus's theory of proportion. It is also widely conjectured that the invention of the axiomatic method was instigated by the inadequacies of the Pythagorian system.

The classic modern example is the discovery of contradiction in the theory of sets during the last decade of the 19th century and first decade of the 20th century (Burali-Forti, B. Russell *et al.*). Remedial concepts formed the basis of new philosophies of mathematics, viz. Intuitionism, Formalism and Logicism. Ultimately, the axiomatic method in its new and formal form was used to limit the conceptual extent of the theory of sets in such manners as to prove the chosen method of most workers in Foundations, not only to avoid the contradictions but, more important, to investigate the status of such frequently used tools as the Axiom of Choice and the Continuum Hypothesis. The success of the latter is exemplied in the modern theory of sets and the achievements of the modern field of mathematical logic. (See the discussion of case 13 in Chapter VI above.)

In stating law 13, we are not referring to those minor cases of contradiction which may be found in special theorems or numerical computations. Thus, "proofs" of the Four Color Theorem preceding the recently acknowledged correct proof (see Appel and Haken, 1977) and "proofs" of Fermat's Last Theorem which have been found inadequate, do not constitute faults in the current conceptual structure of mathematics so far as consistency is concerned. Some have conjectured that the failure to find a proof of the Fermat theorem may reflect an inadequacy in our basic assumptions concerning the integers, but not an inconsistency, which is our concern here. Moreover, special cases where a new conceptual structure has been proposed and investigated, but found to harbor contradiction, may not lead to further investigation.[12]

14. Revolutions may occur in the metaphysics, symbolism and methodology of mathematics, but not in the core of mathematics[13]

This statement requires clarification of the terms used, especially "revolution." Crowe (*loc. cit.*) stipulated that "a necessary characteristic of a revolution is that some previously existing entity (be it king, constitution, theory, terminology or other) must be overthrown and irreversibly discarded." But what about theories that have been abandoned because they have been found to harbor contradictions? For #14 to hold, such occurrences must not be considered as "revolutions." Indeed, one can argue that an inconsistent theory has no place in mathematics, even though its defects are not at first discovered and the theory meanwhile seems to qualify for admission to the core of mathematics. We recognize that we deal here with a debatable opinion; but the choice of what we mean by "revolution" is, of course, subject to the general arbitrariness allowable in the act of defining. We, therefore, exercise our right to exclude theories abandoned because of their inconsistencies. Candidates for "revolution" would be those theories for which alternatives have been published.

An example is the theory of dimension. Here alternative theories have been offered, e.g. that of Fréchet, the Brouwer–Menger–Urysohn

---

[12] For an entertaining history of a fundamental theorem (Euler's formula for polyhedrals) and deficiencies and corrections in proofs thereof, see Lakatos, 1976.

[13] For this law as well as some of the discussion below, see Crowe, 1975a: law 10. Apparently Crowe's motivation in stating this law was (at least in part) due to the wish to show that Kuhn's theory (Kuhn, 1970) was not in this respect applicable to mathematical evolution.

dimension, and homological or cohomological definitions of dimension. None of these has been discarded, although it may be fair to say that the homological (or cohomological) definition is that most commonly used. Nevertheless, all these dimension theories continue to exist, even in the sense that each is still utilized or subjected to further investigation, albeit perhaps only rarely (as in the case of the Fréchet dimension).

Similarly, although we now have an alternative (non-standard analysis) to the classical analysis as developed by the theory of limits, no one expects the latter to be discarded. In geometry, new geometries are added, although the old ones are not discarded. Unlike the natural sciences, in which various theories concerning the natural world can be discarded in the light of new experimental evidence and hence occasion revolutions, mathematics is not bound by experimental verification. Of course, the *user* of mathematics may discard mathematical theories, but this has no reference to the core of mathematics; in *applied* mathematics, revolutions may occur.

The situation is also quite different in the metaphysics of mathematics. For example, until the discovery of the possibility of non-Euclidean geometries, it was considered that Euclidean geometry was a description of the physical world, a sort of natural science. The mathematical culture of today has discarded this belief. Similar observations may be made of such a notion as "the absolute truth of mathematics," usually exemplified by the statement "$2 + 2 = 4$." It is certainly true that 2 apples added to 2 apples yields 4 apples; and this is the case in all applications of the rule. But in today's mathematical culture, the rule has no more status of "truth" than any other mathematical relation.

Also subject to revolution are standards of rigor in mathematical proof. It is notorious in the mathematical culture that standards of rigor have undergone revolutionary changes throughout the evolution of mathematics. Results obtained by 17th- and 18th-century mathematicians through the indiscriminate use of infinite processes, especially infinite series, form virtually a mathematical miracle — as well as testimony to the remarkable intuitions of their creators. Similarly, an ancient Greek mathematician would have accepted without question that a line joining a point in the exterior of a circle to a point interior to the circle would have a point in common with the periphery of the circle; but a 20th-century mathematician would consider that this would have to be proved on the basis of an acceptable system of axioms.

Even appeals to the traditional logic for the proofs of such theorems as the Fundamental Theorem of Algebra[14] and, in general, for existence proofs in general, were challenged by the Intuitionists, headed by L. E. J. Brouwer (1881–1966), who insisted that only such entities as are constructively[15] proved to exist can claim mathematical existence. This represented an extremity of rigor not acceptable to most mathematicians, who continue to use the *reductio ad absurdum* type of proof even for existence theorems. Of course, for purely aesthetic as well as for practical reasons, most mathematicians prefer constructive types of proof when feasible, although they usually take longer to carry out.

Symbolism, too, is subject to rejection and hence "revolution," since frequently old symbols are discarded for new. This, of course, includes names; who, today, would use the term "analysis situs" for "topology," for instance?

Changes in methodology also occur. Some maintain that the introduction by Descartes of algebra into the investigation of geometric forms constituted a revolution in mathematics, in that from that point on analytic methods predominated. However, this was only a methodological innovation, and did not *replace* the former synthetic methods which continued to be employed. It was not even a revolution in methodology in the sense in which "revolution" was defined above. Furthermore, as we have pointed out before, the essence of this matter was that a consolidation of algebra and geometry occurred; the methodological effect was chiefly to *add* to the old methods.

15. **The continued evolution of mathematics is accompanied by an increase in rigor. Each generation of mathematicians finds it necessary to justify (or reject) the hidden assumptions made by previous generations (see II-10)**

Mathematical history bears out this assertion — which assuredly would be agreed to by most mathematicians. After the discussion in II-10, it seems

---

[14]That is, an algebraic equation of degree $n \geq 1$ has exactly n roots in the domain of the complex number system.

[15]Criteria for constructiveness, such as the use only of finite methods, naturally have to be given and hence lead to varying degrees of mathematical existence by constructivity.

hardly necessary to comment further upon it. It should prove interesting to see the effects of future developments in the theory of sets and in mathematical logic, for instance.

16. A mathematical system evolves only through greater abstraction, aided by generalization and consolidation, and usually prompted by hereditary stress (see III-6, 7)

Insofar as abstraction is concerned, this law has been discussed in II-6. As pointed out there, it applies to cultural systems other than mathematical, such as religious and political systems. It seems, indeed, to hold for the evolution of biological systems, if "abstraction" is replaced by "complexity."

So far as mathematics is concerned, the reason for this seems to lie in the fact that the forces which achieve the growth, viz. generalization and consolidation (often accompanied by diffusion), operate principally through the introduction of greater abstraction. The force which prompts the growth is generally hereditary stress, as pointed out in Chapter IV — although there can be cases where the prompting can come from environmental stress.

17. The individual mathematician cannot do otherwise than preserve his contact with the mathematical culture stream; he is not only limited by the state of its development and the tools which it has devised, but he must accommodate to those concepts which have reached a state where they are ready for synthesis (see Wilder, 1953: 439)

This law was stated, not in the form of a law but as an observation in the article cited. The examples given there may be referred to, but we might add the classical example of Boole and mathematical logic. It would have been hardly possible to create mathematical logic in the earlier states of the mathematical culture system; the new attitudes toward algebra which regarded algebraic symbols as not necessarily representing numbers, but as representative of arbitrary objects of thought satisfying certain operational laws was a required forerunner of the creation of symbolic logic.

18.   Periodically, mathematicians assert that their subject is nearly "worked out;" that all essential results have been obtained, and that all that remains is for details to be filled in

We have already commented on this briefly in IV-1. On a cultural level, an example is to be found toward the end of the 18th century. At that time, there seems to have prevailed a feeling that mathematics was becoming "worked out." For some disussion of this, the reader may be referred to Struik, 1948: 198–199.

Expressions of similar feelings on the part of individuals can be found in incidental remarks recorded here and there in the histories of mathematics. For instance, as we remarked in IV-1, Babbage in 1813 asserted that "The golden age of mathematical literature is undoubtedly past." In Wilder, 1953: 440, I recalled the instance of a young mathematician who had just received a Ph.D in topology, but had decided not to work further in the field since "it was obviously all worked out" — this was in the early 1920s, when topology was on the verge of remarkable growth!

19.   Cultural intuition maintains that every concept, every theory, has a beginning (see Wilder, 1953: 428)

Testimony for #19 is provided by the custom of naming theorems, methods, concepts and the like for their supposed originators — many of whom, later historical research shows, were preceded by earlier creators. This is hardly a serious matter, except for those who place a high value on priority, since (as we have pointed out before) the convenience of having a distinguished *name* can be considered to outweigh the loss of historical accuracy involved — especially when the supposed true facts are on record.

The basic difficulty is, of course, due to the general continuity of the cultural process, which oftentimes renders it virtually impossible either to select an originator, or to survey the entire kaleidoscope of cultural events. The common practice of assigning the "beginning" of Greek deductive geometry is a good case in point. Even if such an individual as Thales ever lived, to credit him with the beginning of what must have been a gradual cultural evolution seems absurd.

20. The ultimate foundation of mathematics is the cultural intuition of the mathematical community

In the discussion of law 9, we have already mentioned the cultural intuition. It is, of course, not a fixed thing, but evolves as mathematics itself evolves; nor is it a universally shared entity, since workers in different portions of mathematics have their own intuitions derived from their acquaintance with their special subjects of research.

The attempts to set up a fixed basis for all of mathematics, as was tried during the early part of the present century, constituted notable contributions to mathematical thinking, but as foundations for all of mathematics were doomed to failure. This was not because of such events as the incompleteness theorem of Gödel, often cited as the destroyer of the Hilbert program, but because no structure which will succeed in trapping all the possible concepts present in the intuition, much less those yet unborn, can be derived from the collective intuition. However, mathematics can never escape the necessity of including intuitive concepts among its basic principles, no matter how abstract it may become. Furthermore, without the collective intuitition mathematical research would become sterile; one's intuition is ever a source for new concepts.

21. As mathematics evolves, hidden assumptions are unearthed and made explicit, resulting in general acceptance, or partial or full rejection; acceptance usually follows analysis of the assumption and its justification by newer methods of proof

An excellent example in modern mathematics is the Axiom of Choice. As a hidden assumption it was used even by Cantor. It surfaced first in 1890 with its statement and application by Peano. In 1902 Beppo Levi recognized that it was an independent proof principle, but not much attention seems to have been paid to it until Zermelo published his proof of the well-ordering theorem. Borel promptly pointed out that the two — well-ordering and Axiom of Choice — were equivalent. The significance of the Axiom now being fully realized, further investigation revealed a host of equivalent properties. Fully esconsed in the theory of sets, it is generally accepted except by those adhering to a constructive philosophy of mathematics.

Another outstanding example from history is the emergence of continuity of lines and the real number system needed as far back as Euclid's *Elements*, but unstated and finally evolving in the works of 18th-century researchers (Cauchy, Bolzano, Weierstrass). Other examples can be found in the evolution of theories of convergence and divergence of infinite series, existence of limits, differentials, etc.

22. **A necessary and sufficient condition for the emergence of a period of great activity in mathematics is the presence of a suitable cultural climate, including opportunity, incentive (e.g., emergence of a new field, or occurrence of paradox or contradiction) and materials (see Wilder, 1950: 264)**

For support of law 22 we can refer to the article cited, where its essence was first stated. Its intent is to place responsibility for mathematical innovations on the host culture including, of course, the mathematical culture itself. As we have argued earlier, the availability of the individual talent to initiate and maintain the activity may be taken for granted.

23. **Because of its cultural basis, there is no such thing as the absolute in mathematics; there is only the relative (see Wilder, 1950: 269)**

For justification of law 23 we again refer to the article cited, where it first appeared. The essence of the argument there is the referral of each mathematical concept to the cultural basis which engendered the mathematical structure to which it belongs.

In concluding this chapter, we remark that the above laws are, of course, open to further discussion in regard to their validity. We feel that they are justified in view of the discussions in earlier chapters, but like most "laws" they undoubtedly have exceptions. Possibly qualifications which escaped us should be introduced in some of them.

CHAPTER VIII

Mathematics in the 20th Century; its Role and Future

"The time has come," the Walrus said,
"To talk of many things."

C. L. Dodgson

Although the preceding chapters have already made clear, we hope, the general character of our conception of mathematics, it is now possible, in the light of what has already been said, to be more precise.

1. The place of mathematics in 20th-century cultures

In the first place, we call mathematics a "subculture" of the general culture, and in itself a "cultural system." We have discussed the terms "culture" and "cultural system" in Chapter I. But the world is full of cultures, not all of which have a subculture that we call "mathematics." The so-called "primitive" cultures have number words and today many of them, due to diffusion from the "civilized" cultures, have adopted new counting systems adequate for all their own purposes as well as for their dealings with other cultures. Clearly we have not had in mind such cultures in our previous discourse. Rather we have intended by "general culture" the network of those modern cultures which not only have a well-developed mathematics, but that participate in the on-going evolution of mathematics through their own researches. These cultures, lumped together, can be called a "culture" with as much justice as one can speak of the "United States of America culture," which is actually an aggregate of subcultures between which there is not much more commonality of custom than between the countries of their origins. Due to the modern miracles of communication, we can speak of "the mathematical culture" even though the cultures which serve as its carrier have, in other aspects of life, distinct differences. Presumably no one would speak of "the English–Italian culture," but so far as mathematics is concerned, both English and Italian mathematics are, except for superficial differences, such as possible national preferences for types of mathematics studied, part of the same mathematical culture.

As time passes, the mathematical culture spreads by diffusion into the cultures of nations that have theretofore not been active in mathematical research. Now (1979) that the Chinese culture is establishing relations with countries active in mathematics, one can expect to see, barring adverse political change, a resumption and flowering of mathematical work in China. Even more interesting to observe, however, will be the advances made in mathematics in the so-called "developing countries," particularly those on the African continent (exclusive of Egypt and South Africa, which already form part of the mathematical culture). These latter countries, so long excluded from participation in scientific development of the modern world, will undoubtedly commence, through diffusion, to set up scientific and in particular mathematical subcultures. Already individuals from such countries have appeared in European and American universities and although many of them may elect not to return to their homes, ultimately the requisite libraries and laboratories will be built in their home countries that will entice them to return.

2.   Future "dark ages?"

The foregoing remarks are, of course, contingent upon favorable environmental stresses. The political and demographic problems currently clouding the general world picture do not augur well in this regard, so that it is not possible to predict with any certainty regarding the spread of mathematics and science. History records the "dark ages" during which the great works of the Greek and Hellenistic cultures lay virtually dormant, to be revived in the Western European cultures with great difficulty. There is certainly no guarantee that another "dark age" will not again occur. What this would do to mathematics in its present highly evolved state can only be conjectured. As a subculture of the general culture, environmental stresses acting upon it from the general culture would presumably be catastrophic. We must not forget, however, that when we speak of a "dark age" we are referring only to the "general culture" as defined above. The historical "dark age" which encompassed Western European civilization did not adversely influence the Arabic cultures. The latter, indeed, experienced a flowering of arts and letters, and although they did not produce the mathematics that their successors in Western Europe later produced, they kept alive Greek and Babylonian mathematical works; for this, the present-day mathematical culture can be grateful. One might hope, therefore, that a future "dark age" affecting the present general culture, would also be paralleled by cultures which, like the Arabic, would preserve the mathematical works that would presumably be lost during the decline of the general culture. Naturally, this is all conjectural; a rejuvenation of modern mathematics after such a dark age would encounter great difficulties, due to the advanced abstract character of modern mathematics, with which the rejuvenation of Greek mathematics after the historical Dark Age could hardly compare.

Much of the Greek mathematical work was irretrievably lost, some of it due to the destruction of the Alexandrian Library. Some of this material, as, for example, Apollonius's later books on conic sections, some Renaissance scholars tried to restore. But other works, such as Euclid's on *Porisms* and *Surface Loci,* were lost. However, has modern mathematics suffered from the loss of any important concepts that might have been contained in these works? It is quite doubtful that this could be the case, considering the higher level of abstraction that has been attained by modern mathematics. It is possible, of course, that some geometrical work considered new (such as the theorem later discovered by Morley) was contained in some of this lost work. But the general overall evolution of modern mathematics has probably been little affected, if at all, by such losses.

On the other hand, a catastrophic loss, or loss by a future dark age, would be of much greater import to the general evolution of culture. Such a loss would undoubtedly be of a quite different nature than that brought on by the destruction of Greek mathematics. Consider the state of the world today. The finiteness of our planet is much more in evidence today. At the time of the Greek and Hellenistic civilizations, no one in these cultures was aware of the Mayan culture and vice versa. Today we are not only cognizant of cultural developments throughout our planet, but even the most "primitive" of cultures must have some awareness of the existence of our Western culture even if it is only from the occasional planes that traverse their skies. Moreover, instead of there being a single library of the importance of the Alexandrian Library in the Hellenistic age,[1] great libraries today dot a substantial area of the world map. It is difficult to conceive of a dark age, or even a catastrophe that would destroy all these libraries. But what good are libraries if no one can read and understand their contents? Presumably a "dark age" would imply such a situation.

3. The role of mathematics in the 20th century

We have already observed above that what we call the "mathematical culture" is at present generally restricted to the "advanced" cultures of the world. There are also, generally speaking, the areas that have become technologically proficient. Mathematics affects, and is used by, the technological parts of these cultures as much as by their sciences. One of the latest gifts from mathematics to the general, as well as to the technological, parts of culture is the computer. The computer has not only speeded up the evolution of technology in modern society, but bids fare to revolutionize much of its way of life, as did the autmobile and airplane.

However, except for so-called computer theory, the computer has now passed, as did its predecessors, the operational parts of arithmetic and the abacus, into the hands of engineers and commercial enterprises in whose hands the cited evolution is being carried out. What about other products of modern mathematics? It becomes more apparent, year by year, that the core of mathematics serves the general culture by producing new concepts whose future, in addition to their uses in mathematics proper, will be to move into the general culture, meeting needs unforeseen at present.

A good example of the latter is the theory of groups. In its origins most abstract and hardly of any foreseeable use outside the core of mathematics, during the present century it has played a crucial role in physics. It is recalled by Dyson, 1964:

> In 1910 the mathematician Oswald Veblen and the physicist James Jeans were discussing the reforms of the mathematical curriculum at Princeton University. "We may as well cut out group theory," said Jeans. "That is a subject which will never be of any use in physics." It is not recorded whether Veblen disputed Jeans's point, or whether he argued for the retention of group theory on purely mathematical grounds. All we know is that group theory continued to be taught. And Veblen's disregard for Jeans's advice turned out to be of some importance to the history of science at

[1]Even in the Hellenistic age, there were important private libraries to be sure, but they were restricted to the Greek, Roman and other Near Eastern areas.

Princeton. By an irony of fate group theory later grew into one of the central themes of physics, and it now dominates the thinking of all of us who are struggling to understand the fundamental particles of nature. It also happened by chance that Hermann Weyl and Eugene P. Wigner, who pioneered the group-theoretical point of view in physics from the 1920's to the present, were both Princeton professors.

This anecdote can be duplicated in essence by many others. Who, during the latter part of the 19th century and the first decade of the 20th, would have foreseen that the studies initiated by Boole, Frege, Russell and Whitehead, and Hilbert would soon generate notions such as recursiveness (Gödel, Turing, etc.) which would serve as one of the basic notions of computer theory? Or that matrix theory, initiated by Cayley (*ca.* 1858), would turn out to be the precise tool needed by Heisenberg in 1925 for mathematical description of quantum mechanics phenomena? Or that the theory of analytic functions, involving complex quantities, would turn out to have widespread applications in physics and especially electrical phenomena? One could go on with the citing of such cases. The "moral" should, however, by now be clear: That the "pure" mathematics of today will be the applied mathematics of tomorrow. As Braithwaite observed: "...pure mathematicians should be endowed fifty years ahead of scientists" (Braithwaite, 1960: 49). (Cf. also the quotation from von Neumann in II-6; and on the general relationship between mathematics and the sciences, see Wilder, 1973.)

As remarked earlier, during our discussion of environmental influences on mathematics, specifically the influences of governmental and private grants for research, it was inevitable that such granting agencies would try to direct the course of research. This could be done most effectively by refusing grants for research that was not considered "mission-oriented,"[2] even though the intent may be only to fund research that may prove useful to the granting agency, not to try to *direct* research in general. The United States Air Force Office of Scientific Research took such action over a decade ago.

From a cultural point of view, one can question from two standpoints the feasibility of trying to direct the course of evolution of a cultural system: (1) Whether it is possible, (2) whether, granting it to be possible, any individual or organization has the prophetic wisdom to choose the right

---

[2]By "mission-oriented" is meant research whose results have probable applications to the technology or aims of the granting agency.

course to accomplish the desired goals. Throughout our discussion above, we have studied the ways in which certain processes and forces affect the evolution of a cultural system. Among these we have admitted the influence of individual action, as the synthesizer of cultural forces impinging on the individual consciousness.

As mathematics evolves the individual mathematician caught in the culture stream, not only borrows his ideas therefrom, but is affected by the forces inherent therein as conveyed to him by his fellow mathematicians. In the Veblen–Jeans anecdote quoted above, Veblen, a "pure" mathematician and a trained axiomatizer, was no doubt already aware of the unifying character of group theory and, possibly, of the way in which concepts from the core of mathematics had effected physical and philosophical thinking. Jeans, on the other hand, was well versed in the uses of classical mathematics in physics and astronomy. But in 1910, group theory was not only not yet "classical," but it represented a type of mathematics — structural in nature rather than algorithmic — which at that time was undoubtedly considered as bordering on the philosophical. From a knowledge of the background and training of the two men — Veblen and Jeans — one might accurately have predicted the respective attitudes they would take toward the maintenance of group theory in the undergraduate curriculum. Unfortunately most universities did not show the foresight exhibited by Princeton; courses in group theory in the United States universities did not become common until the late 1920s or early 1930s, at which time the importance of group theory for both modern algebra and applications to other sciences began to force their introduction into the curricula.

Returning to the more general question of whether it is possible to direct the course of evolution of a cultural system, the answer depends upon deciding who or what, if anything, does the directing. Certainly both people and organizations — carriers of culture — participate in and do the work of the evolving system. But can they — individuals or organizations — anticipate the likelihood of future events and take action which will change the course of evolution? Here, as in the concept of culture, anthropologists disagree. Some remark, "Everyone agrees that wars should not be allowed to occur; but measures to prevent war have proved futile. Wars do keep occurring, and seem part of an uncontrollable aspect of culture." Others disagree. After the last World War, many idealists, physicists in particular, formed organizations to "outlaw" the use of atomic weapons. The futility of such actions seems apparent. Other attempts were, and are being, made to prevent the use and construction of nuclear plants for producing energy. But if a culture must have energy in order to function, will it not find it without regard for the feelings of individuals?

So far as the question of whether any individual or organization has the prophetic wisdom to choose the right course to accomplish desirable goals (e.g. elimination of war) is concerned, the answer seems unequivocally negative. One can cite instance after instance from the history of mankind of undeniably wrong decisions concerning the accomplishment of desired goals. And insofar as the matter of what constitutes desirable objectives for a cultural system are concerned, past history seems only to indicate the futility of trying to decide upon such objectives.

Despite these pessimistic observations — which may be taken for what they seem to be worth — the study of cultural systems does have *predictive* value. This, in short, is what we have tried to show in our statements of "laws" in Chapter VII. Observation of the behavior of cultural systems permits the indulgence in such predictions with some degree of success (as we have already stated in the preliminary material of Chapter VII). It is much the same as the case of the zoologist whose special interest is the behavior of animals; given his knowledge of the past behavior of a special type of animal, he can state rather accurately how such a creature will react to a given set of circumstances. Conversely, then, if one wishes to produce such reactions, he will continue to set up the kind of conditions that he knows has produced them.

If, then, it is desired to stimulate a cultural system to the production of certain ends, it may be possible to set up conditions which will produce them — within, however, strict limitations. The attempts by governmental agencies in the United States to stimulate increased scientific production during and following World War II are an example of this. The desired response from the scientific culture, and in particular the mathematical cultural system, was produced. But this is a far cry from directing a system to produce only certain kinds of mathematics; such attempts seem doomed to failure. The course of mathematical evolution, as we have observed above, seems always to have tended to greater abstraction; in law 16 of Chapter VII we stated: "A mathematical system evolves only through greater abstraction, aided by generalization and consolidation, and usually prompted by hereditary stress." To attempt to turn mathematics to the exclusive servicing of so-called "practical" needs would only doom it to sterility.

Certainly environmental stresses from the host culture, for instance, have an effect which does influence mathematical evolution. At the present time (1979) in the United States, teaching and academic positions for which many Ph.D.s have been trained are becoming scarce due to decreasing numbers of students and inflationary developments which make it difficult to afford academic training. One consequence of this is that many Ph.D.s turn to government and industrial positions which seemingly are becoming more available due to the growing technology. Many of these positions demand some knowledge of computer theory, and accordingly mathematics students are being urged to augment their training in core mathematics with courses in computer theory and practice. Obviously such developments can, and most likely will, affect the evolution of core mathematics. But one must note that *it is a result of cultural forces acting upon the mathematical cultural system, not the direction dictated by any individual or organization bent upon changing the course of mathematical evolution.*

4. The uses of mathematics in the natural and social sciences

Everyone is familiar with the uses of mathematics in the ordinary affairs of life; these have been with us ever since the days of Sumer and Babylon and constitute the substance of arithmetic as universally taught in the elementary schools. It is not such uses, however, that concern us here. Rather we refer to the kinds of mathematics that have been created in modern times, usually for "mathematics' sake," and that constitute the major part of the core of mathematics. We have already mentioned the theory of groups above and its uses in physics.

The uses of mathematics in the other sciences are generally of two kinds: (1) As a tool and (2) as a source of conceptual configurations. It is the "tool" or, as it is sometimes called, the "language" uses of mathematics that we have just referred to in the preceding paragraph. It is also those uses regarding which those vexing persons who are eternally inquiring of a mathematical researcher's work, "Of what use is this?" have reference. The answer expected is of the type that Faraday unfortunately had to give according to the (apochryphal?) story in which he was asked the same question concerning his work on magnetism: "Some day you'll be able to tax it!"

In recent times the "tool" uses of mathematics have increased enormously, not only in the applications of computer theory, but in the uses of such subjects as mathematical logic, finite mathematics, combinatorics, etc. Such uses are analogous to those made of calculus and classical analysis by the early astronomers, physicists, etc. These are understandably important and have affected the course of mathematical evolution, especially through the demands made by the sciences for special tools. These matters are generally well recognized today.

Not so well known are the uses cited in (2) regarding mathematics as a source of conceptual configurations. These need to be more generally recognized, since they call attention to the importance of core mathematics. Mathematics created in response to the request for a new conceptual tool is not, generally speaking, likely to prove fruitful so far as the evolution of mathematics is concerned. Rather it is mathematics which has been created in response to hereditary stress, "mathematics for mathematics' sake," which is most likely to further the advance of mathematics and — it should never be forgotten — leads to conceptual structures that will cut a wide swath in future scientific thinking. For Einstein, the Ricci calculus and Riemannian geometry stood ready for his needs, so that even as tools the core mathematics proves itself in meeting future demands. But most theoretical physicists, for instance, know that when their imagination seems to fail them, they can turn to the core of mathematics for further insights. To quote Dyson (*loc. cit.*) again, "for a physicist mathematics is not just a tool by means of which phenomena can be calculated; it is the main source of concepts and principles by means of which new theories can be created."

In II-12 we cited a "problem" that many physicists and philosophers have pondered over and which the Nobel prize winner Wigner calls "the unreasonable effectiveness of mathematics in the natural sciences," citing in particular the use of complex numbers in quantum mechanics. But Wigner also called attention to the manner in which mathematically formulated laws gave meaningful results when applied to situations which extended far beyond the originally intended domains of application. He went on to affirm what he called *the empirical law of epistemology,* namely "the appropriateness and accuracy of the mathematical formulation of the laws of nature in terms chosen for their manipulability, the 'laws of nature' being of almost fantastic accuracy but of strictly limited scope."

In his conclusion Wigner states: "The miracle of the appropriateness of the language of mathematics for the formulation of the laws of physics is a wonderful gift which we neither understand nor deserve. We should be grateful for it and hope that it will remain valid in future research and that it will extend, for better or for worse, to our pleasure even though perhaps also to our bafflement, to wide branches of learning."

Whether similar observations could be made regarding the role of mathematics in the social sciences awaits the results of future research regarding human behavior. The intimate relationship that existed between mathematics and physics throughout history is not duplicated in the case of the social sciences, so that the "miracle" of which Wigner speaks may well not occur in this case.

On the other hand, the social sciences, compared to physics, are only on the threshold of their evolution; and there is evidence of a growing relationship between them and mathematics, especially in the case of such older disciplines as economics. It may well turn out that this relationship will eventually produce a situation not unlike that which exists today between mathematics and the natural sciences.

Virtually throughout the evolution of mathematics there have existed forces of varying degrees of intensity, affecting mathematics from both without and within. During the earlier stages of mathematical evolution, hereditary stress was almost undetectable. This was due principally to the fact that mathematics had not yet reached the status of a cultural system, although despite this (as we have earlier remarked) traces of the operation of hereditary stress are detectable in Babylonian mathematics.[3] As one approached the modern era, with mathematics gradually achieving the status of a sub-culture, the operation of hereditary stress helped to transform mathematics into a self-sufficient cultural system possessing its own interests and generating its own internal forces such as symbolization, consolidation, abstraction, etc. At the same time, environmental stress of a cultural nature in the form of demands from the growing natural sciences continued to affect the way in which mathematics grew. For a time,

[3]Regarding the operation of cultural forces on mathematics in Babylon, see EMC: Chap. 1, §4.

particularly in the 16th–19th centuries, these two evolving entities, mathematics and the natural sciences, were often represented in the same person. It was not until the splits due to the growing complexity of both mathematics and the various natural sciences, that the "specialist" appeared and was formally baptized with the growing institutionalization of science as "mathematician," "physicist," "chemist," etc.

With this turn of events, some mathematicians became "mathematical physicists," "engineers," "statisticians," "actuaries," and, when attached to commercial institutions, "applied mathematicians." In terms of cultural systems, what was happening was the impinging of the environmental cultural vectors on the mathematical vectors of the host cultures, resulting in the attraction of potential "pure mathematicians" into the environmental specialties. Despite this turn of events, the internal factors continued to grow, and core mathematics continued to expand not only due to internal pressures but from conceptual structures suggested by the cultural environment. The latter were usually not of immediate use to the environment, but in most cases were destined to become so after being successfully developed within mathematics; such was the case with statistics, probability, and, later, computing theory.

The general pattern seems to have been for the core mathematics to develop, along with its more abstract interests, special structures which turn out to have "application" to cultural developments external to mathematics; we have already touched upon this phenomenon in II-12 ("A Problem"). In the opposition between environmental stress and hereditary stress, there continually results a diffusion from the core mathematics to the environment. This is being exemplified, in recent times, in the creation of such subjects as game theory, computer theory, operations research, linear programming, systems analysis, communications science, optimization, etc. This has resulted in a transformation of what one means by the term "applied mathematics." Before World War II, applied mathematics consisted mainly of classical analysis and its applications to mechanics and physics — generally speaking to the natural sciences and engineering. Such applications have been generally absorbed into the specialties that use them, and in some cases, "applied mathematics departments" have rejoined the original core departments. The new "applied mathematics," consisting of such subjects as those named above, are a "different breed," and involve modern algebra more than the classical analysis. Today mathematicians are to be found working under strange labels in various kinds of technological concerns such as the aircraft industry, brokerage houses, banks, etc. Some of these mathematicians, one suspects, never solved a differential equation, but are adept at applying finite mathematics and especially mathematical methods, to social and commercial problems.

In this situation, it is not surprising that environmental pressures on mathematics departments have arisen to adjust or augment their curricula to take account of the new "applications." And it is not surprising that many mathematicians, especially those with little historical perspective, imagine that they are facing a novel situation and wondering how to cope with it. But it is not new; early in the 20th century, in response to analogous situations, "applied departments" split off from the core departments, although others were able to handle the "crisis" by curricula and staff adjustments. "To split or not to split" in order to solve the problem requires careful study of the local situation and the effects on the mathematical personalities involved. As we have emphasized in law 16 of Chapter VII, and in the relevant discussion in II-6, the greatest power in mathematics is achieved through increased abstraction, and to attempt to interrupt this process would be fatal to the core mathematics upon which all applications feed. But with the accumulation of new concepts there arise the need of interpretation for the "user" of mathematics as well as curriculum adjustments that must be made to instruct the latter.[4] How the individual department will handle the matter will depend a good deal on the breadth of knowledge which those in charge have regarding how mathematics evolves and has evolved, and what forces have been increasingly involved in the process. Especially will a knowledge of similar developments in the early part of the century help in making such decisions.

5. Relevance to historiography

We have used historical episodes, such as those recounted in Chapter III, to arrive at certain conceptions concerning the cultural nature of mathematics. In concluding this work, we mention some of the opinions we have formed in regard to the writing of history — historiography. At the

[4]Throughout this discussion we have been influenced by the situation in the United States. However, with advancing technology and industrialization similar problems will certainly arise, if they have not already arisen, in the "developed" nations.

present time, historical writing appears to have quite adequately covered, insofar as materials are available, the events which have made up the history of mathematics. As new sources have been uncovered, history has continued to expand; such was notably the case about a half century ago when Neugebauer added materially to our hitherto meager knowledge of Babylonian mathematics through his translations of Babylonian texts hidden in the mass of baked clay tablets unearthed by "Assyriologists" and others.

Certain portions of the history of mathematics have of necessity been treated from an impersonal point of view, simply because their individual creators were unknown. With the advent of the Greek period, at first certain individuals such as Thales are made responsible for the beginnings of geometry, although much doubt has been cast upon their ability to invent the concepts credited to them. Later workers, such as Eudoxus and Plato, appear historically credible. Moreover, knowledge of the existence of incommensurables and the Zeno paradoxes furnishes a credible cultural impulse toward the ensuing results achieved by these workers; mathematical historians generally recognize this. Even the matter of the introduction of the axiomatic method and its accompanying logical apparatus has been investigated by various scholars, notably Szabo (*loc. cit.*) with a view to ascertaining the reasons therefor.

As one approaches the more modern period, when the individual creators of mathematical concepts can be identified, there seems to be a growing tendency to resort to a chronological account of individual persons and their achievements. Paradoxically, this parallels the growth of mathematics toward cultural status, so that cultural factors could be brought in to further enlighten the course being taken by mathematics. As we have emphasized, mathematics has not grown in a vacuum. Its roots are imbedded in the needs of the cultures that fostered it — a fact historians have recognized, to be sure. But as it achieved cultural status, new forces have emerged to complement those imposed by the environment. To ignore these is to only partially understand how mathematics has evolved. Even during the formative stages, in addition to the environmental stresses impinging on it, mathematics was affected by cultural lag and resistance, recognized social forces that can impede the diffusion of desirable notions from one culture to another — presumably a factor in the Greek failure to profit adequately from Babylonian developments.

With the achievement of cultural status, such internal forces as hereditary stress, consolidation, selection, symbolization and abstraction have played an increasingly important part in the evolution of mathematical thought. To ignore these forces and the manner in which they have influenced characteristic patterns of evolution seems to deprive both the historian and the students who follow him of a fuller understanding of the historical process. Of course, mathematics is done by individuals, but these individuals share a common, albeit variable and diverse, mathematical culture, and along with the study of the achievements of a Gauss or Riemann, one should study the culture, both mathematical and environmental, in which they lived and worked, in order to achieve a fuller understanding and appreciation of what they accomplished. The complaint of a Bolyai: "One can hardly hold it possible that two, even three people knowing nothing of one another, and almost simultaneously and moreover by different paths, should almost completely have settled the matter," would no longer be heard if knowledge of the predominance of multiple invention were widespread and the reasons therefor understood.[5] Familiarity with the patterns described above and especially of the laws of Chapter VII, would seem to explain many occurrences in the history of mathematics that are generally not well understood.

For example, there has been some feeling that identifying the forces of mathematical evolution as we did in EMC is unproductive, in that all that has been accomplished thereby has been to give *names* to processes already recognized. Aside from the fact that naming a concept of itself confers more status as well as greater recognition of its importance, we hope that the analysis contained in Chapters IV and V of hereditary stress and consolidation have added to the comprehension of these forces. Perhaps we should also have devoted more study to environmental stress; but in this case, one can expect the concept to be already rather well known and appreciated.

The evolution of mathematics has formed a cultural stream, tenuous at times but nonetheless always resuming its flow, from which the individual mathematician chooses an area or areas of interest from among those

---

[5]Similar to the complaint of Bolyai just quoted, the author has a vivid memory of almost identical complaints being made by his mentor, R. L. Moore, regarding work published simultaneously with his own. Since my knowledge of cultural matters was nil at the time, I could not give Moore any aid to understanding of the situation.

existing at the time, and makes his contributions — after which the stream flows on to his successors. Each such event, depending upon its importance, the historian records. But as an event in a cultural system, it deserves further treatment regarding the reasons for its occurrence at that particular time and place, particularly regarding what antecedent and current cultural forces rendered it possible.

APPENDIX

Footnote for the Aspiring Mathematician

We hope that we have, in this book, made clear what we think needs recognition in this advanced age, namely, that mathematics is a subculture, created by man and yet forming a cultural system that in a certain sense dominates its carriers, who can at best observe its laws such as we have enumerated in Chapter VII. It would seem appropriate, then, to list certain items of an ordinary nature for the student who hopes to make mathematics a career; like all advice, they can be accepted or rejected. We have in mind particularly those students who are about to commence research in mathematics.

1. Don't worry too much about being *first* to obtain a result. Not only is it bad for your nervous system, but most "firsts" in mathematical history were really not firsts. Moreover, if it is a good result, showing that you are capable of doing important research and it is intended as your dissertation for the Ph.D., then your mentor, if he has any sense, will see that you get your degree. Never forget that we all "stand on the shoulders of giants."[1] I have never asked a seasoned research mathematician if he had ever been anticipated by someone else in his work, and received a negative answer.

2. In the same vein, always be aware that when one makes a great breakthrough such as in the form of a solution of an outstanding problem or in the creation of an important new concept, other mathematicians may

---

[1]In case you think Issac Newton was the first to enunciate the idea embodied in this quotation, read R. K. Merton's book *On the Shoulders of Giants,* N.Y., Harcourt, Brace & World, 1965, a scholarly, witty and charming essay

independently and simultaneously duplucate it. But don't resent this; it is the way in which our culture works.[2]

3. Don't worry that your research, if it is in the core of mathematics, is too abstract. Remember that "The paradox is now fully established that the utmost abstractions are the true weapons with which to control our thoughts of concrete fact" (Whitehead, 1933: 48. See III-6, 7 above).

4. Be always aware that when the going becomes too difficult in your area of research, there probably lurk ideas in neighboring parts of mathematics, whose consolidation with the concepts with which you are working may hold the key to solving your difficulties. Mathematics has an underlying unity, to which every field of mathematics as it approaches being "worked out" may appeal.

5. Always remember that since you are contributing, through your synthesis and creations, to the mathematical cultural stream, it will be advisable to maintain contact with what your fellow mathematicians are doing and thinking. With modern means of communication through meetings, journals, and advanced technologies for recording and transmitting new results, it is easy to do this. In this way you will preserve contact with your mathematical culture and establish a two-way partnership that will help both you and your mathematical colleagues to maintain interest in the work you are doing. Furthermore, you will be made aware not only of the problems most needing solution, but perceive when concepts have reached a state when they are ready for synthesis.

6. Never fear that your field of interest is becoming "played out." If you really think this is the case, by all means recall what was stated in IV-1 and in connection with law 18 of Chapter VII above; look about you (mathematically) and the chances are almost certain that you will find a way of consolidating your interests with more general interests; or, as happens when the physical scientist looks to mathematics for new conceptual structures, you will become aware of structures in other parts of mathematics which will suggest new structures in your own field of interest.

7. Never forget the famous unsolved problems. Their solution probably only awaits the time when the relevant parts of mathematics will have

---

[2]Perhaps English natural scientists have a way of avoiding such situations through a cultural accommodation. According to Watson, 1969: 19, it "would not look right" for a scientist in England to begin work on a problem that he knew was already being worked on by someone else in England.

generated concepts and methods that will make their solutions possible. Recall the famous Four Color conjecture, whose solution finally became possible when the computer was finally built that made possible checking some 1200 or so cases that had to be eliminated.

8. You are starting at a higher level than your forebears, unencumbered by outdated methods and possessing tools that they did not have. As a consequence you will gradually find that you have powers that enable you to handle concepts that would often be unintelligible to those who did the groundwork in your field.

Bibliography

Appel, K. and Haken, W. (1977) "The solution of the four-color-map problem," *Scientific American,* vol. 237, pp. 108–121.

Banach, S. and Tarski, A. (1924) "Sur la décomposition des ensembles de points en parties respectivement congruentes," *Fundamenta Mathematicae,* vol. 6, pp. 244–277.

Bell, E. T. (1945) *The Development of Mathematics,* 2nd ed., New York, McGraw-Hill.

Bell, E. T. (1951) *Mathematics: Queen and Servant of Science,* New York, McGraw-Hill.

Beth, E. W. (1959) *The Foundations of Mathematics,* Amsterdam, North-Holland Publishing Co.

Boyer, C. B. (1968) *A History of Mathematics,* New York, John Wiley & Sons.

Bourbaki, N. (1960) *Éléments d'Histoire des Mathématiques,* Paris, Hermann.

Braithwaite, R. B. (1960) *Scientific Explanation,* New York, Harper and Row Publishers.

Brouwer, L. E. J. (1976) *Collected Works,* 2 volumes, ed. H. Freudenthal, Amsterdam, North-Holland Publishing Co., or New York, American Elsevier Publishing Co.

Browder, F. E. (Ed.) (1976) *Mathematical Developments Arising from Hilbert Problems,* Providence, R.I., American Mathematical Society.

Bunge, M. (1962) *Intuition and Science,* Englewood Cliffs, N.J., Prentice-Hall, Inc.

Carr, G. S. (1970) *Formulas and Theorems in Pure Mathematics,* 2nd ed., New York, Chelsea. [The 1886 ed. was titled "A Synopsis of Elementary Results in Pure Mathematics."]

Childe, V. G. (1946) *What Happened in History,* New York, Penguin Books.

Cole, J. R. and Cole, S. C. (1973) *Social Stratification in Science,* Chicago, University of Chicago Press.

Conant, L. L. (1896) *The Number Concept,* New York, Macmillan.

Coolidge, J. L. (1934) "The rise and fall of projective geometry," *American Mathematical Monthly,* vol. 41, pp. 217–228.

Coolidge, J. L. (1963) *A History of Geometrical Methods,* New York, Dover. (Originally published in 1940 by Oxford University Press.)

Court, N. A. (1954) "Desargues and his strange theorem," *Scripta Mathematica,* vol. 20, pp. 5–13, 155–164.

Coxeter, H. S. M. and Greitzer, S. L. (1967) *Geometry Revisited,* New York, Random House/Singer.

Crowe, M. J. (1975a) "Ten 'laws' concerning patterns of change in the history of mathematics," *Historia Mathematica,* vol. 2, pp. 161–166.

Crowe, M. J. (1975b) "Ten 'laws' concerning conceptual change in mathematics," *ibid.,* pp. 469–470.

Crowe, M. J. (1978) [A review of EMC_1], *ibid.,* vol. 5, pp. 99–105.

Dantzig, T. (1954) *Number, the Language of Science,* 4th ed., New York, Macmillan.

Dieudonné, J. (1975) "Introductory remarks on algebra, topology, and analysis," *Historia Mathematica,* vol. 2, pp. 537–548.

Dresden, A. (1924) "Brouwer's contributions to the foundations of mathematics," *Bulletin of the American Mathematical Society,* vol. 30, pp. 31–40.

Dubbey, J. M. (1978) *The Mathematical World of Charles Babbage,* Cambridge Univ. Press.

Dyson, F. J. (1964) "Mathematics in the physical sciences," *Scientific American,* vol. 211, pp. 127–146; reprinted in *Mathematics in the Modern World,* San Francisco, W. H. Freeman, 1968, pp. 249–257.

Feit, W. and Thompson, J. G. (1963) "Solvability of groups of odd order," *Pacific Journal of Mathematics,* vol. 13, pp. 775–1029.

Gilfillan, S. C. (1971) *Supplement to the Sociology of Invention,* San Fransisco, San Fransisco Press.

Grattan-Guiness, I. (1971) "Towards a biography of Georg Cantor," *Annals of Science,* vol. 27, pp. 345–391.

Hadamard, J. (1949) *The Psychology of Invention in the Mathematical Field,* Princeton, N.J., Princeton University Press.

Heath, T. L. (1956) *The Thirteen Books of Euclid's Elements,* 2nd ed., Cambridge, England, The University Press; republished in paperback in 3 volumes, New York, Dover Publications, Inc.

Henle, J. M. and Kleinberg, E. M. (1979) *Infinitesimal Calculus,* Cambridge, Mass., MIT Press.

Heyting, A. (1956) *Intuitionism. An Introduction,* Amsterdam, North-Holland Publishing Co.

Hilbert, D. (1901–2) "Mathematical problems," English translation from original German, *Bulletin of the American Mathematical Society,* vol. 8, pp. 437–479.

Iltis, H. (1966) *Life of Mendel,* translated by E. and C. Paul, New York, Hafner.

Johnson, D. M. (1977) "Prelude to dimension theory: The geometrical investigations of Bernard Bolzano," *Archive for History of Exact Sciences,* vol. 17, pp. 261–295.

Klee, V. (1979) "Some unsolved problems in plane geometry," *Mathematics Magazine,* vol. 52, pp. 131–145.

Kline, M. (1953) *Mathematics in Western Culture,* New York, Oxford University Press.

Kline, M. (1972) *Mathematical Thought from Ancient to Modern Times,* New York, Oxford University Press.

Koppelman, E. (1975) "Progress in mathematics," *Historia Mathematica,* vol. 2, pp. 457–463.

Kroeber, A. L. (1917) "The superorganic," *American Anthropologist,* vol. 19, pp. 163–213.
(1944) *Configurations of Culture Growth,* Berkeley, University of California Press; reprinted 1969.
(1948) *Anthropology,* New York, Harcourt, Brace, rev. ed.
(1952) *The Nature of Culture,* Chicago, University of Chicago Press.

Kuhn, T. S. (1970) *The Structure of Scientific Revolutions,* 2nd ed., Chicago, University of Chicago Press.

Lakatos, I. (1976) *Proofs and Refutations,* ed. J. Worral and E. Zahar, Cambridge, University Press. Originally published in *British Journal for the Philosophy of Science,* vol. 14, 1963–4, pp. 1–25, 120–159, 221–243, 296–342.

Lewis, J. A. O. (1966) *Evolution of the Logistic Thesis in Mathematics,* Ann Arbor, University of Michigan Dissertation.

Lusin, N. (1930) *Lecons sur les Ensembles Analytiques et leurs Applications,* Paris, Gauthier-Villars.

Menninger, K. (1954) *Number Words and Number Symbols,* Cambridge, Mass., MIT Press.

Merton, R. K. (1973) *The Sociology of Science,* Chicago, University of Chicago Press.

Monk, J. D. (1970) "On the foundations of set theory," *American Mathematical Monthly,* vol. 77, pp. 703–711.

Moore, E. H. (1910) *Introduction to a Form of General Analysis*, The New Haven Mathematical Colloquium, New Haven, Yale University Press.

Moore, R. L. (1932) *Foundations of Point Set Theory*, Providence, R.I., American Mathematical Society Colloquium Publications, vol. 13; rev. ed. 1962.

Neugebauer, O. (1957) *The Exact Sciences in Antiquity*, Providence, R.I., Brown University Press.

Oakley, C. O. and Baker, J. C. (1978) "The Morley trisector theorem," *American Mathematical Monthly*, vol. 85, pp. 737–745.

Poincaré, H. (1946) *The Foundations of Science*, translated by G. B. Halsted, Lancaster, Penn., Science Press.

Poncelet, J.-V. (1865) *Traité des Propriétés Projectives des Figures*, 2ᵉ ed., Paris, Gauthier-Villars, 2 vols.

Poudra, M. (1864) *Oeuvres de Desargues*, Paris, Leiber, 2 vols.

Price, D. J. de S. (1961) *Science since Babylon*, New Haven, Yale University Press.

Ramanujan, S. (1962) *Collected Works*, ed. G. H. Hardy, P. V. Seshu Aiyar and B. M. Wilson, New York, Chelsea Publishing Co. [Original published by Cambridge University Press in 1927.]

Robinson, A. (1966) *Non-Standard Analysis*, Amsterdam, North-Holland Publishing Co.

Riemann, G. F. B. (1854) "Uber die Hypothesen welche der Geometrie zu Grunde Liegen," Göttingen, Habilitationsschrift.

Rodin, M., Michaelson, K. and Britan, G. M. (1978) "Systems theory in anthropology," *current Anthropology*, vol. 19, pp. 747–762.

Rudin, M. E. (1975) *Lectures on Set Theoretic Topology*, Providence, R.I., American Mathematical Society [No. 23 of Regional Conference Series in Mathematics.]

Russell, B. (1937) *The Principles of Mathematics*, 2nd ed., New York, W. W. Norton.

Sanchez, G. I. (1961) *Arithmetic in Maya*, Austin, Texas, published by the author (2201 Scenic Dr.).

Sarton, G. (1950) "Query no. 129. Desargues in Japan," *Isis*, vol. 41, pp. 300–301.

Shapiro (1970) *Aspects of Culture*, Freeport, N.Y., Books for Libraries Press.

Schoenflies, A. (1908) *Die Entwickelung der Lehre von den Punktmannigfaltigkeiten*, II, Leipzig, Teubner.

Simpson, G. G. (1952) *The Meaning of Evolution*, New Haven, Yale University Press.

Struik, D. J. (1948) *A Concise History of Mathematics*, 2 vols., New York, Dover.

Swinden, B. A. (1950) "Geometry and Girard Desargues," *Mathematical Gazette*, vol. 34, pp. 253–260.

Szabo, A. (1964) "The transformation of mathematics into deductive science and the beginnings of its foundation on definitions and axioms," *Scripta Mathematica*, vol. 27, pp. 27–48A, 113–139.

Taton, R. (1951a) *L'Oeuvre Mathématique de G. Desargues*, Paris, Presses Universitaires de France. Contains original text of the *Brouillon Projet* based on the text found by M. Pierre Moisy in the Bibliothèque National *ca.* 1950.

Taton, R. (1951b) *L'Oeuvre Scientifique de Monge*, Paris, Presses Universitares de France.

Taton, R. (1951c) *La Géométrie Projective en France de Desargues à Poncelet, Conférence fait au Palais de la Découverte le Février 1951*, Université de Paris.

Taton, R. (1960) *Les Origines de la Géométrie Projective*, in Notes du 2ᵉ Symposium International d'Histoire des Sciences (Pisa-Vinci, 16–18 June, 1958).

Thompson, E. S. W. (1977) *Sociocultural Systems: An Introduction to the Structure of Contemporary Models*, Dubuque, Iowa, Wm. C. Brown, Publ.

Tuckerman, B. (1971) "The 24th Mersenne Prime," *Proceedings of the National Academy of Sciences, U.S.A.,* vol. 68, pp. 2319-2320.

Tylor, E. B. (1958) *The Origins of Culture,* New York, Harper Torchbooks.

Veblen, O. (1921) *Analysis Situs,* New York, American Mathematical Society Colloquium Publications, vol. 5, part 2. 2nd ed., 1931.

Von Neumann, J. (1961) "The role of mathematics in the sciences and society," in *Collected Works,* ed. A. H. Taub, New York, Pergamon Press, vol. 6, pt. 477–490.

Watson, J. D. (1969) *The Double Helix,* New York, New American Library, a Signet Book.

White, L. A. (1947) "The locus of mathematical reality," *Philosophy of Science,* vol. 14, pp. 289–303; republished in somewhat altered form as Chapter 10 of White, 1949.

White, L. A. (1949) *The Science of Culture: A Study of Man and Civilization,* New York, Farrar, Straus.

White, L. A. (1959) "The concept of evolution in cultural anthropolopgy," in *Evolution and Anthropology: A Centennial Appraisal,* Washington, D.C., The Anthropological Society of Washington.

White, L. A. (1975) *The Concept of Cultural Systems, A Key to Understanding Tribes and Nations,* New York, Columbia University Press.

Whitehead, A. N. (1933) *Science and the Modern World,* Cambridge, England. Reprinted in paperback, New York, Pelican Mentor Book, 1948. The New American Library.

Whyte, L. L. (1950) "Simultaneous discovery," *Harper's Magazine,* vol. 200, pp. 23–26.

Wiener, Chr. (1884) *Lehrbruch der Darstellenden Geometrie,* Leipzig, Teubner, 2 vols. (Vol. 2 was published in 1887).

Wigner, E. P. (1960) "The unreasonable effectiveness of mathematics in the physical sciences," *Communications on Pure and Applied Mathematics,* vol. 13; reprinted several times as, for example, in T. L. Saaty and F. J. Weyl (eds.), *The Spirit and Uses of the Mathematical Sciences,* pp. 123–140, New York, McGraw-Hill, 1969.

Wilder, R. L. (1932) "Point sets in three and higher dimensions and their investigation by means of a unified analysis situs," *Bulletin of the American Mathematical Society,* vol. 38, pp. 649–692.

Wilder, R. L. (1950) "The cultural basis of mathematics," *Proceedings International Congress of Mathematicians,* vol. 1, pp. 258–271.

Wilder, R. L. (1953) "The origin and growth of mathematical concepts," *Bulletin of the American Mathematical Society,* vol. 59, pp. 423–448.

Wilder, R. L. (1959) "The nature of modern mathematics," *Michigan Alumnus Quarterly Review,* vol. 55, pp. 302–312.

Wilder, R. L. (1965) *Introduction to the Foundations of Mathematics,* 2nd ed., New York, John Wiley & Sons.

Wilder, R. L. (1967) "The role of intuition," *Science,* vol. 156, pp. 605–610.

Wilder, R. L. (1968) *Evolution of Mathematical Concepts, An Elementary Study,* New York, John Wiley & Sons (referred to in text by "EMC").

Wilder, R. L. (1974) *Ibid.,* London, Transworld Publishers Ltd., Transworld Library Edition in paperback.

Wilder, R. L. (1978) *Ibid.,* Milton Keynes, England, Open University Press.

Wilder, R. L. (1969) "Trends and social implications of research," *Bulletin of the American Mathematical Society,* vol. 75, pp. 891–906.

Wilder, R. L. (1973a) "Mathematical rigor, Relativity of standards of," in *Dictionary of the History of Ideas,* New York, Charles Scribner's Sons, 4 vols.; see vol. 3, pp. 170-177.

Wilder, R. L. (1973b) "Mathematics and its relations to other disciplines," *Mathematics Teacher,* vol. 66, pp. 679–685.

Wilder, R. L. (1974) "Hereditary stress as a cultural force in mathematics," *Historia Mathematica*, vol. 1, pp. 29–46.

Wilder, R. L. (1979) "Some comments on M. J. Crowe's review of Evolution of Mathematical Concepts," *Historia Mathematica*, vol. 6, pp. 57–62.

Zaslavsky, C. (1973) *Africa Counts*, Boston, Prindle, Weber and Schmidt.

Zirkle, C. (1951) "Gregor Mendel and his precursors," *Isis*, vol. 42, pp. 97–104.

Index